BOLD
FORECAST

BOLD FORECAST

The Hurricane Agnes deluge

GARY LETCHER

FOREWORD
DR. GREG FORBES

PENN DEL
PRESS

AUTHOR'S NOTE

This is a true story, told in the style of a novel. It is based upon exhaustive research and interviews with many of the persons involved. In some instances the sequence of events is adjusted to make the narrative easier to follow. Verbatim dialog is cited as such. True names are used for public officials, Weather Service personnel, volunteer observers, first responders, and flood victims. The names of other persons have been modified to protect their privacy.

Cover and interior design by Bea Reis Custodio
@beareiscustodio
London

ISBN 979-8-9862267-0-5 (Paperback)
ISBN 979-8-9862267-1-2 (eBook)

PENN DEL PRESS

CONTENTS

The Flood (original art by Diane Bild, 2021)

FOREWORD

Dr. Greg Forbes

There are just a few weather events in peoples' lives that they remember forever – like Hurricane Katrina and its flooding of New Orleans. For those living in parts of the Northeast, Hurricane Agnes in 1972 was one of those kinds of events. Over 210,000 people had to flee their homes from the massive flooding, and 122 were killed. Countless others had some first-hand encounter. I was one of the latter.

My own experience with Hurricane Agnes was not as devastating as for hundreds of thousands of others. I had just completed my undergraduate studies at Penn State University and was at home in Latrobe in western Pennsylvania during the few days between the end of final exam week and commencement. But flooding from Agnes kept me and most others from getting back to University Park for that ceremony. I tried to drive there, but didn't even get ten miles before a washed-out road blocked my path. Later in the summer I went back to pick up my diploma, a Bachelors of Science Degree in Meteorology. I could easily see changes in the familiar landscape resulting from the raging Agnes floodwaters. The Little Juniata River had destroyed the old one-lane bridge on Rt. 45 at the tiny community of Spruce Creek – not far from where President Jimmy Carter would later go fishing. Along an alternate route, the normally tiny Bald Eagle Creek had obviously turned into a small river. Even the tiny streams in State College, normally not even noticed flowing beneath small bridges along Atherton Street, had made this main thoroughfare through town impassible.

But this was minor compared to massive flooding on the Susquehanna River, winding its way across northeastern, central, and southeastern Pennsylvania, through Wilkes-Barre, Berwick, Bloomsburg, Sunbury, Selinsgrove, Harrisburg, and dozens of communities along the way, and by other rivers and streams in Pennsylvania and New York. Waters rose twelve to nearly nineteen feet above flood stage on the Susquehanna, West Branch Susquehanna, and Schuylkill Rivers in Pennsylvania, and on the Chemung River in New York, setting records at most locations that have not since been exceeded. In all, long-standing flood records were eclipsed in six states. All told, Agnes caused $3.5 billion in damage back in 1972 which, after adjusting for inflation, was the costliest United States hurricane until Andrew in 1992.

Agnes was not that noteworthy while it was a hurricane, making landfall in Florida on June 19 as Category 1 on the Saffir-Simpson Scale. It weakened to a tropical depression as it moved across Georgia and North Carolina, but unexpectedly strengthened back to a tropical storm over Virginia. From this time on, Agnes became a deadly rainstorm, as it linked up with a cold, slow-moving upper-air low and stalled over New York and Pennsylvania. Some places in this hilly terrain got eighteen inches of rain over several days, with 14.8 inches falling in one 24-hour period in Schuylkill County, Pennsylvania.

Agnes showed that, as many tropical storms and hurricanes have done in more recent years, the leading cause of death from these systems is usually from inland flooding. Agnes also showed – as did Tropical Storm Allison in Houston, Texas in 2001 – that when it comes to flood potential, tropical storms and weak hurricanes can cause as much or more flooding than strong hurricanes.

For readers too young to have lived through Agnes, Gary Letcher does an excellent job of turning back the clock to 1972, presenting the way of life during this period and recreating the mood that prevailed before Agnes quickly grabbed attention away from the Vietnam War and Watergate. He vividly portrays the impacts that the flood had on the lives of those who lost everything. The book weaves in the history of this flood-prone area, and efforts that had been taken to reduce the risk from floods. Letcher describes how rivers are measured and forecast, how warnings are conveyed. But these sections are not a dry compilation of facts. Instead, he makes this information interesting by blending it into stories of the daily lives, activities, and personal experiences of citizens, emergency management officials, volunteer weather observers, and National Weather Service hydrologists leading up to, during, and after the flood.

There are lessons to be learned about floods and flood forecasting by just about everyone from reading *Bold Forecast*. Heavy rains in 2004, 2006 and 2011 caused residents along the Susquehanna to look back at Agnes and pray that it wouldn't get that bad again. Residents of the Midwest learned the power of floodwaters in 1993 and 2008. The lesson learned is that an Agnes-like flood could happen again, and along any river in the United States.

ACKNOWLEDGMENTS AND APPRECIATION

Many and sincere thanks to the following persons for their time and gracious assistance in preparation of this work:

Kevin Hlywiak and Scott Kroczynski, hydrologists, who scoured the archives of the Middle Atlantic River Forecast Center for records of the flood, instructed the author on the science of river forecasting, and illuminated operations of the Forecast Center then and now.

The late Myron "Mike" Gwinner, hydrologist, Middle Atlantic River Forecast Center (1972), who for many hours shared his recollections about the Agnes floods and the contemporaneous operations and personnel of the RFC.

Dr. Bob Kuligowski, National Oceanic and Atmospheric Administration, Satellite Meteorology and Climatology Division, who provided access to his comprehensive collection of publications and documents pertaining to the Agnes floods, and who reviewed the manuscript for scientific and administrative accuracy.

John Comey, executive assistant to the Director, Pennsylvania Emergency Management Agency, formerly Public Information Officer to Pennsylvania's Council on Civil Defense (1972), who advised about operations and personalities of the Council on Civil Defense involved in the Agnes response.

The following additional persons provided valuable contributions in the form of interviews, recollections, and documents.

Middle Atlantic River Forecast Center hydrologists:
Peter Ahnert, Hydrologist-in-Charge
Joseph Ostrowski
Dan Zanzalari
Ned Pryor
Lars Feese (1972)
Michael Mark (1972)
National Weather Service:
Albert Kachic, Chief Regional Hydrologist, Eastern Region (1972)

John Chiaramonte, meteorologist and forecaster, Syracuse/
  Binghamton (1972)

Keith Eagleston, Senior Climatologist, Northeast Regional
  Climate Center

Dr. Ted Letcher, Atmospheric Scientist, United States Army
  Corps of Engineers Cold Regions Research and Engineering
  Laboratory.

Fred Kepner, assistant superintendent, Wilkes-Barre,
  Pennsylvania, Public Schools (1972).

Dr. Robert Brill, Director of Science, Corning Museum of
  Glass (1972).

Janet McCormick and Freddie Jones, step-daughter and step-
  son of the late Ola. D. White, Hydrologist-in-Charge, Middle
  Atlantic River Forecast Center (1972).

Wayne Vanderpool, volunteer observer of the Susquehanna
  River at Towanda.

Marsha Field, daughter of the late LaVern Root, volunteer
  observer of the Susquehanna River at Towanda (1972)

Judy Souchik, and the late Olga Souchik, daughter and wife
  of the late Nicholas Souchik, volunteer observer of the
  Susquehanna River at Wilkes-Barre and executive director of
  the Luzerne County Council on Civil Defense (1972).

* * *

I also want to thank editors Michelle Rubin at *Writers Ally* and
Joyce Finn at *Writers and Critters* for their firm and expert hands in
improving the manuscript, Shirley Letcher for meticulously proof-
reading the final draft, Diane Bild for her original art, Courtenay
Kling for assistance with images, and Bea Reis Custodio for her
outstanding design of the interior and cover of this book.

Last, and perhaps most important, special thanks to my wife
Shirley, and my late mother Sherrie Letcher, for encouraging me,
over many years, to persist in writing this book. Agnes is a story
that deserves to be told, and would not have been told here but for
their persistence.

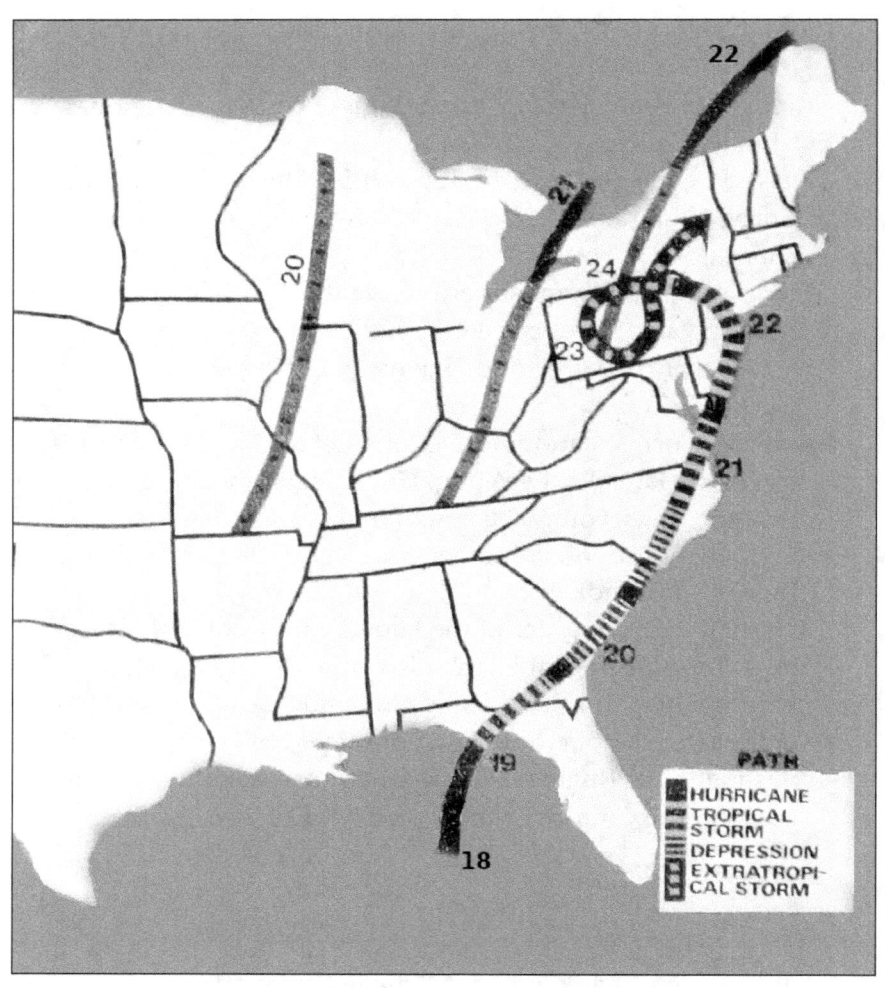

The path of Agnes, showing interaction with a cold front approaching from the Midwest (*U.S. Army Corps of Engineers,* The Corps Responds)

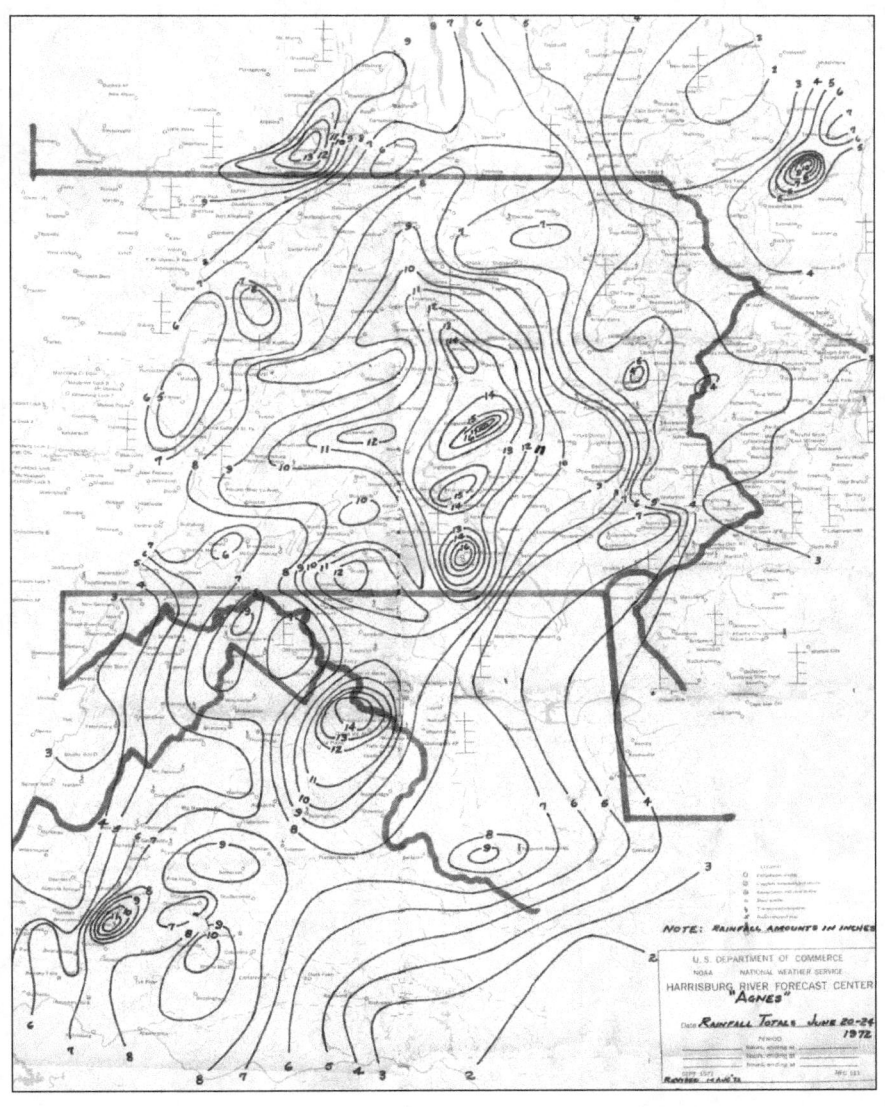

Contemporaneous hand-rendered rainfall contour map, with 10+ inches of rain over five states and a maximum 18.2 inches near Harrisburg, Pennsylvania (NWS, River Forecast Center)

NEVER AGAIN!

August 15, 1936
Wilkes-Barre, Pennsylvania

Twenty-year-old Nicholas Souchik rose from his seat on the curb, pulling his girlfriend Olga up with him. They'd waited in the sweltering sun for hours, but Nick and Olga didn't mind. After all, it wasn't every day the President of the United States paid a visit.

"Here he comes, here he comes!" The crowd lining Wilkes-Barre's Public Square surged in anticipation as President Franklin Roosevelt's motorcade drew near. Nick and Olga leaned out and craned their necks for a better view up Market Street. To Nick Souchik, who like many in Wilkes-Barre never ventured far from home, Roosevelt's visit was the most excitement he had ever known.

Nick and Olga were but two of the half million residents of Pennsylvania's Wyoming Valley, and beyond, who turned out to see the chief executive.[1] Never before had such a huge crowd assembled in Wilkes-Barre. This was United Mine Workers country, and Roosevelt's pro-labor policies made him popular here. More to the point, residents looked to the president for protection against ravaging floods of the Susquehanna River. The president came, in his words, to "look-see" construction of protective dikes beginning to rise above the river's banks.

On this day the river registered only two feet on Wilkes-Barre's official gauge, near the record low, so shallow in some places a person could wade across. The new levees seemed massive overkill against this placid stream.

Five months earlier the Susquehanna was anything but placid. The harsh winter of 1935-36 blanketed the hills of the Mid-Atlantic and New England states with four feet of snow, and ice on the river grew two feet thick. River watchers knew the spring thaw could bring terrible flooding. On February 18 the Weather Bureau warned *the next few weeks will probably bring the crisis.*[2]

The crisis came right on schedule. Four inches of rain in mid-March deluged the watershed as temperatures rose into the 50s. The rain and melting snow couldn't soak in where the soil was frozen, but ran directly into the river and its tributaries. The Susquehanna rose inexorably, carrying a burden of broken ice that plowed into bridges and piled up at bends. The Luzerne County engineer begged the War Department to drop bombs on the ice jams so the river could flow freely.

There was ultimately no protection from the rising waters. The river crawled over its banks on March 17 and extended its icy fingers, four feet deep, through downtown Wilkes-Barre and the surrounding communities. They called it the St. Patrick's Day flood.[3]

The Susquehanna crested at thirty-three feet on Wilkes-Barre's river gauge, higher than ever before. The frigid waters swamped the communities of Wilkes-Barre and Hanover on the east side, and Forty Fort, Plymouth and Kingston on the west. Sheets of dirty ice were left to melt in city streets, and thick, rancid mud penetrated thousands of riverside homes and businesses.

Rivers rose to record levels across the Middle-Atlantic region and into New England; Pittsburgh, Johnstown and the Connecticut Valley were especially hard hit. Scores of people died and thousands were left homeless. Property damages came to $250 million,[4] an enormous sum in Depression-era America.

Souchik's home in Edwardsville stood above the flood and suffered no harm, but he spent many days clearing icy mud and debris from the homes of relatives not so fortunate. Souchik had never witnessed a flood before, but learned to ever after take the Susquehanna very, very seriously.

The federal government was powerless to help victims of the floods. 1936 was long before establishment of the Federal Emergency Management Agency, FEMA. Instead, the Red Cross and other civic and church organizations stood at the front lines of the disaster. The President could only urge donations to charity to aid the victims. And, to be sure, several hundred young men of Roosevelt's Civilian

Conservation Corps – CCC boys, they were called – came to help with the cleanup.

The St. Patrick's Day flood rallied the government to action. The Flood Control Act of 1936 catapulted through Congress only three months after the waters receded. One of the great pieces of New Deal legislation, the law for the first time recognized that "flood control ... is a proper activity of the Federal Government."[5] From now on, Uncle Sam would foot at least part of the bill. Indeed, an appropriation under the new law paid for the dikes Roosevelt came to look-see.

How high should the new walls be? Flooding had come to the Wyoming Valley before. Many in the crowd remembered the freshet of 1902, and a handful of old-timers recalled a big one back in 1865. Both times the Susquehanna rose almost as high as St. Patrick's Day. Could the river rise even higher? The Army Corps of Engineers said yes: in a worst-case, the Susquehanna at Wilkes-Barre could climb to thirty-six feet, three feet higher than the record reached on St. Patrick's Day. The new dikes would have to be high enough to hold back even the greatest possible flood.

Now, this August day, the president's special Packard touring car, surrounded by a phalanx of secret service, passed no more than fifteen feet in front of Nick and Olga. The crowd cheered as the beaming Roosevelt waved his panama hat high over his head. "Hello!" the president shouted simply, over and over, "Hello!" The National Guard fired blanks in salute, and every child on Public Square waved a small American flag. Pandemonium.

Nick ran behind the motorcade as Olga clutched his hand. If Roosevelt was going to make a speech, Nick didn't want to miss it. In fact, the President didn't plan to deliver remarks at Wilkes-Barre. His visit was highly choreographed, a show for the voters. 1936 was an election year, and it couldn't hurt to remind the locals exactly who was building their dikes. Roosevelt arrived via train at Scranton, been driven by car through Wilkes-Barre to inspect the levees (and be inspected by the cheering crowds), and at the Market Street station would board the train to his private home at Hyde Park. The entire visit lasted only an hour and a half.

Yet as Nick, Olga, and a half-million like them cheered, the president was moved by their tenacity and optimism. They had suffered, and looked to him for reassurance. Just before boarding his train Roosevelt spoke quietly to officials gathered to see him off. "If I have anything to do with it," he promised, "the Federal government will work with to you to protect you from floods like that again".[6]

Nick Souchik remembered President Roosevelt's visit for the rest of his life. The people of the Wyoming Valley were grateful to the President – grateful for the CCC boys he sent to their aid, grateful for honoring their need and resilience, and, especially, grateful for building the levees and walls that would, as the Wilkes-Barre Times-Leader said, "wipe out forever the menace of the Susquehanna."[7] *A flood like this could never happen again!*

EL NINO

Friday, June 9, 1972
Gulf of Guayaquil, Ecuador

The fishermen of Ecuador's anchovy fleet were having a terrible year. Last season brought record harvests, but now nets cast from the 70-foot boats were often hauled in empty. The fishermen knew their quarry had been driven away by El Nino, an incursion of warm water from the South Pacific that sometimes came at Christmastime (and so named for the Christ-child). Indeed, this was the strongest El Nino since 1925. Surface sea temperatures seven degrees above normal drove the anchovies to cooler water – and influenced the weather across North and South America.[1]

As happened every afternoon, air heated by the warm ocean, cooked even more in the Equatorial sun, rose into the unstable atmosphere. Moisture in the rising air condensed to droplets, then froze into little spheres of ice that climbed ever higher to form towering thunder clouds. The fishermen huddled under their boat's canopy as drenching rains poured over the coastal hillsides and nearby seas.

Such clouds usually rained themselves out in the cool evening, but this cluster of storms grew strong enough to persist after the sun went down. It continued to feed off the warm water of El Nino as it drifted northward, and on June 12 crossed the Isthmus of Panama to the Caribbean Sea. Months later, researchers studying satellite photographs pointed to this tempest in the Pacific as the germ of what was to become Hurricane Agnes.[2]

* * *

Friday, June 9, 1972
Harrisburg, Pennsylvania

Stop the War! A knot of demonstrators hailed passers-by in front of Harrisburg's Federal Building, and a young woman held up a placard bearing the familiar peace sign. The target of their protest was the Selective Service Commission – the agency that administered the hated draft – on the top, eleventh, floor of the building. A civil servant returning from lunch shoved through the protestors, "Hippie freaks, get a job." Moments later, another worker flashed a V with his fingers.

Ola D. White – most people knew him only as O.D. – watched the young protestors from the sidewalk across Walnut Street. White was the Hydrologist-in-Charge at the Middle Atlantic River Forecast Center (RFC), a division of the National Weather Service. The RFC shared the eleventh floor of the Federal Building with the Selective Service Commission.

White crossed the street to return to his office. In his crisp white shirt, wide necktie and heavy black glasses, White looked the epitome of the "establishment" the protestors despised, *the man.* Yet as he sidled through, White offered a curt nod and tight-lipped smile. These kids weren't much different than the students he taught in high school many years before, intelligent and full of promise. The demonstrators paused as White stepped by, and the young woman with the Peace sign returned his smile.

Two months earlier, getting into the Federal Building had not been so easy. Then, demonstrators ringed the building in a human chain, shouting, "Free the Harrisburg Seven!" A group of Catholic priests and nuns were on trial in the fifth-floor courtroom, charged with a conspiracy to kidnap National Security Advisor Henry Kissinger and dynamite steam tunnels under Washington. The court declared a mistrial after the jury deadlocked on the most serious charges.[3]

White rode the elevator up to the RFC and stepped to his office at the southwest corner of the building. A pile of personnel forms, memos from headquarters, and other documents awaited his action. As Hydrologist-in-Charge he spent more time these days on administrative tasks and less on actual science. Enjoying a few moments before plowing into the papers, White lit his ever-present pipe and gazed out the window. At this hour the sun glimmered off the Susquehanna River four blocks to the west, a wide ribbon of silver.

White had come a long way since his youth in Scotts Hill, Tennessee, a poor community in the rolling hills east of Memphis. The area was at the heart of the Tennessee Valley Authority, a 1930s New Deal program to harness the great Tennessee River for electricity and flood control, lifting the region out of Depression poverty. O.D. White grew up knowing the power of a big river.

After the 1941 attack on Pearl Harbor the Navy sought trainees for its meteorology program. White leapt at the opportunity, and was soon forecasting the weather for naval operations in the Pacific. After the war he signed on with the Ohio River Forecast Center in Cincinnati, and was soon promoted to lead the new Middle Atlantic RFC at Harrisburg.

It was up to White, as the first Hydrologist-in-Charge, to organize the network of weather and river observers, set up protocols for receiving data, and refine the algorithms for calculating river flow. White distinguished himself – and enhanced the stature of river forecasting nationwide – when he accurately predicted record floods on the Delaware River during 1955's one-two punch from Hurricanes Connie and Diane.

The most important work of the RFC was, and remains, flood forecasting. Hydrologists determine if rainfall or snow-melt will bring a river over its banks, then alert response authorities and riverside communities. But in addition to flood warnings, river forecasting is essential to businesses and people who use the river every day. Power plant operators depend on the RFC's predictions to control water for cooling their machinery. Water and sewage utilities adjust their intakes and discharges according to predicted river flows. Dam operators need to know how much water is coming so they can maintain reservoir levels. And, recreational users of the river – marinas, boaters and anglers – want to know if there will be enough, or too much, water to carry their boats.

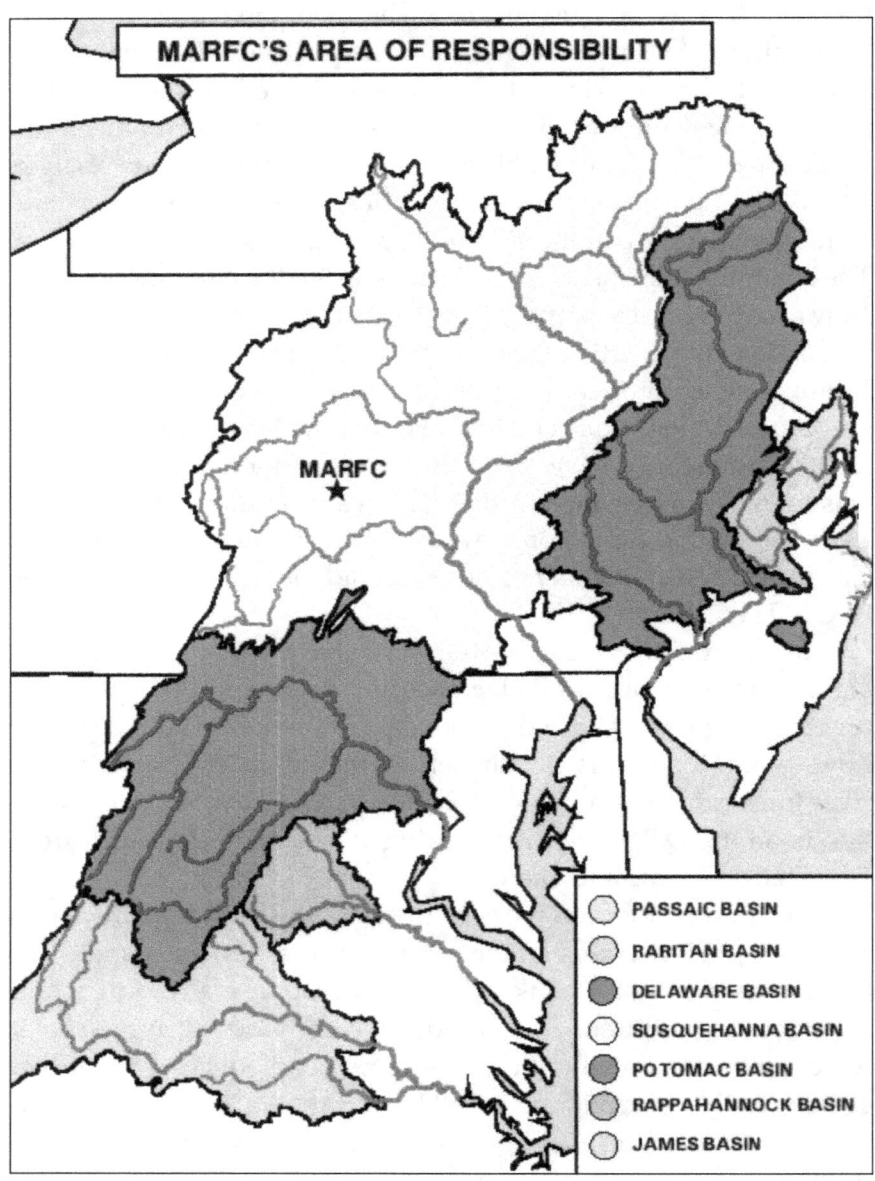

Region covered by the Middle Atlantic River Forecast Center (NWS)

River Forecast Center staff, 1972. Seated from left: O.D. White, Don Close; standing: Nick Pavick, Bruce Whyte, Lars Feese, Paul Marin, Mike Gwinner (U.S. Department of Commerce)

Seven hydrologists and a handful of support staff worked under White's supervision. Most of them kept normal business hours Monday to Friday, while on weekends one of the hydrologists came in to receive data from river observers. Although White managed the office, his hydrologists knew their jobs and needed little scientific supervision. White said he "worked up" to Weather Service headquarters, while the staff hydrologists "worked down" with the network of observers and the "customers" who relied on the RFC's predictions of flow in the region's waterways.

White could see some of those customers from his office window. At the City Marina, a small fleet of sailboats, fishing craft, and cabin cruisers crowded their piers. The marina – on City Island in the middle of the Susquehanna – received the RFC's river predictions every day. A few miles downriver, the nuclear power plant under construction at Three Mile Island promised a safe and inexpensive supply of electricity. The plant's operators would need White's forecasts to regulate the flow of cooling water to the reactors, without which there could be a catastrophic meltdown. Even further below, the Philadelphia Electric Company relied on the RFC's forecasts to store or release water at Conowingo Dam – the largest hydroelectric plant in the eastern United States.

* * *

Ever curious, O.D. White would have been fascinated by the fishermen of Guayaquil. But this June afternoon, in his office thousands of miles away, he gave no thought to their plight. And the fishermen, miserable in the rain and the paucity of their catch, certainly gave no thought to him.

Anchovy boat in the Gulf of Guyaquil, Ecuador (ANIMALS by VISION/Alamy)

RIVER OBSERVER

Saturday, June 10, 1972
Wilkes-Barre, Pennsylania

A glaze of frost coated the grass in Nick Souchik's small backyard. His eyes widened despite the early hour. How could there be frost in June? He rubbed condensation from the pane, better to read the little thermometer mounted outside. Thirty-four degrees. "Damn! That's got to be a record!"

Souchik was keenly aware of the weather. As the volunteer "cooperative observer" of the Susquehanna River at Wilkes-Barre, he checked the stage of the river and weather data first thing every morning. His instruments were located near the courthouse, where Souchik worked as executive director of Luzerne County's Council on Civil Defense. When he finished breakfast he would drive downtown to take his readings, then report his observations to the River Forecast Center at Harrisburg.

Souchik's volunteer role as an observer for the RFC fit hand-in-glove with his job as director of Civil Defense. Planning for disaster response, including floods, was part Civil Defense's mission, so Souchik's intimate knowledge of the river complemented his professional duties. He came to know some of the other observers and the hydrologists at the RFC, and gained at least a superficial understanding of how river forecasting works. Convenient location of the river gauge only yards from his courthouse office made it even easier to mesh his twin positions.

Thirty-four degrees was indeed a record, beating the previous low for the day by an incredible nine degrees. It was, and remains, the coldest temperature ever recorded at Wilkes-Barre in the month of June. The cold came behind a powerful polar front that surged down from Canada the day before, bringing record low temperatures throughout the eastern United States. Souchik would later hear there was a hard freeze in the mountains west of the city, with official readings as low as 27: ice-on-the-birdbath cold, only a week before the first day of summer!

The cold was heralded by violent thunderstorms. As the front advanced, the hot, humid air that stifled the region was pushed up into massive cumulonimbus clouds, their tops stretched into gigantic dark anvils. Thunder echoed off the hills surrounding Wilkes-Barre, looming ever louder as the storm approached. Streetlights came on, their sensors confused by the afternoon darkness. Gusty winds whipped the rain into undulating sheets, and lightning arced from the snarling clouds – in a few places striking homes and businesses.

The storm ended as quickly as it began. The cold front quick-marched to the south, purging the air of the humidity, dust and haze with which it had been laden. Wilkes-Barre's sky that evening was as clear and azure blue as anyone could remember.

* * *

Nick Souchik and his wife Olga lived in Plains, a suburb two miles north of Wilkes-Barre, in a "double block" duplex with a postage-stamp patch of grass behind. Their home had a front porch hard against the sidewalk, where Olga rushed to chat with the neighbors every evening. Many of Wilkes-Barre's neighborhoods had a similar homey, if gritty, appeal. The patina of gray coal dust on white clapboard unmistakably said "Pennsylvania".

The coal industry pervaded the Wyoming Valley. Deposits of anthracite discovered in the nineteenth century brought such wealth that Wilkes-Barre was known as the "Diamond City". The anthracite of the Wyoming Valley was the best coal in the world, and for a hundred years it fired the industries of Philadelphia and warmed the tenements of New York. With their fortunes from the mines, coal barons of the Wyoming Valley built palatial mansions and extravagant public buildings along the Susquehanna. But the reign of King Coal couldn't last forever. Most of the mines closed by 1960, exhausted of their black riches.

With the coal industry had come the miners, hundreds of thousands of immigrants from Russia, Italy, Ireland, Cornwall, and, especially, Poland. The population of Wilkes-Barre was twenty percent Polish, and in the nearby towns of Nanticoke and Plymouth as much as half.[1] *Pierogi* or *kielbasa* were as likely to be served at the tables of the Wyoming Valley as meatloaf or pork chops.

Bound by their common ancestries, their lives in the mines, and their cultural and religious affiliations, the people of the Wyoming Valley watched out for one another. Many families were woven to others by marriage, by neighborhood, or by fraternal and civic organizations. The Masons, Elks, and Kiwanis were huge here. Nick Souchik himself was chief of the Plains Lions Club, and often hosted meetings in his own parlor.

* * *

Souchik plucked his baseball cap from the hook by the door as Olga reminded him they had choir practice at church that morning. She did *not* want to be late. Stepping outside, he deliberately pursed his lips and puffed out, chuckling at the absurdity as his breath made a little cloud in the cold air.

From Souchik's home it was a straight shot down River Road to the courthouse. Along the way fields of corn alternated with copses of woodland. He noted that the corn had already grown knee-high. Souchik wasn't a farmer, but he knew that all the recent rain was good for the crops.

The 7:00 a.m. newscast came on the radio as Souchik turned his car into the vacant parking lot at the courthouse. He listened casually as he steered back toward the little shed that housed the river gauge. First there was news of the war: American bombers continued to pummel industries in North Vietnam. Some of the men in the National Guard, his men, were on active duty and in Vietnam now. In politics, Senator George McGovern campaigned in New York for the presidential primary the following week. Finally, as Souchik braked to a halt, the newscast told of an overnight flood in South Dakota. He turned up the volume and cocked his head. A flash flood at Rapid City had killed dozens, maybe more. Rescue efforts were under way.

Nick Souchik had no idea that the storm in South Dakota, 1,500 miles away, was spawned by the same polar front that frosted his backyard that morning. Still, part of his job at the Civil Defense was planning for flood response, and he knew that his own Susquehanna, though flowing calmly today, could rise over its banks and wreak havoc. He saw the disastrous St. Patricks Day flood back in 1936, and the city had a couple of close calls since. The news from Rapid City underscored the importance of his daily measure of the big river. Thank God, and Roosevelt, for Wilkes-Barre's floodwall!

Souchik left his car and ambled over the grass to the gauge house. The tiny concrete structure, barely five feet square, stood on the banks of the Susquehanna just inside Wilkes-Barre's flood wall. The wall, a corrugated steel fence atop an earthen levee, had been constructed after the 1936 St. Patrick's Day flood to keep the river from ever again rising into the city.

A gap in the wall near the courthouse allowed for passage of North Street to a bridge over the river. If a flood was forecast, Wilkes-Barre's Department of Public Works would hoist huge floodgates into the gap and anchor the gates to the roadway. Public Works practiced every year, and in 1964 put the gates in place for real, when the river climbed nearly to the 1936 record. There was little damage. Wilkes-Barre's wall had been tested, and residents knew they were safe from the worst the Susquehanna could offer.

As Souchik opened the steel door of the gauge house its hinges groaned, rusted from the recent rains. Inside the shed a small shelf held the new "bubble gauge", its digital readout indicating the river "stage" in glowing red numerals.[2] The bubble gauge measured air pressure in a tube that ran under the river, and converted that measurement to an electronic display.

Souchik was surprised to learn, when training as river observer, that the "stage" of the river is not the same as its depth, but rather is the elevation of the river surface above a local "zero" point, or

Nick Souchik, river observer and director of Luzerne County Council on Civil Defense, at a "staff gauge" on the North Street Bridge at Wilkes Barre (courtesy of Wilkes-Barre Times-Leader)

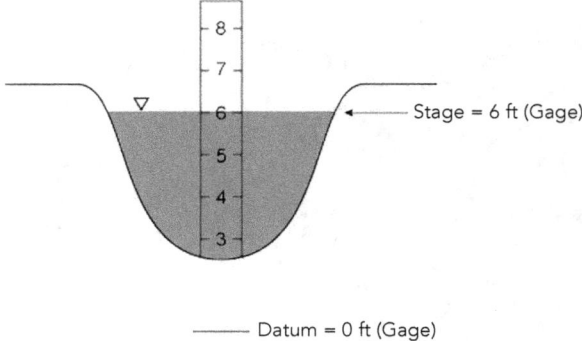

"Stage" is the elevation of the river's surface above an arbitrary "zero" point, or datum. Stage is not the same as "deep" (courtesy of Dr. Brad Finney, Humboldt State University)

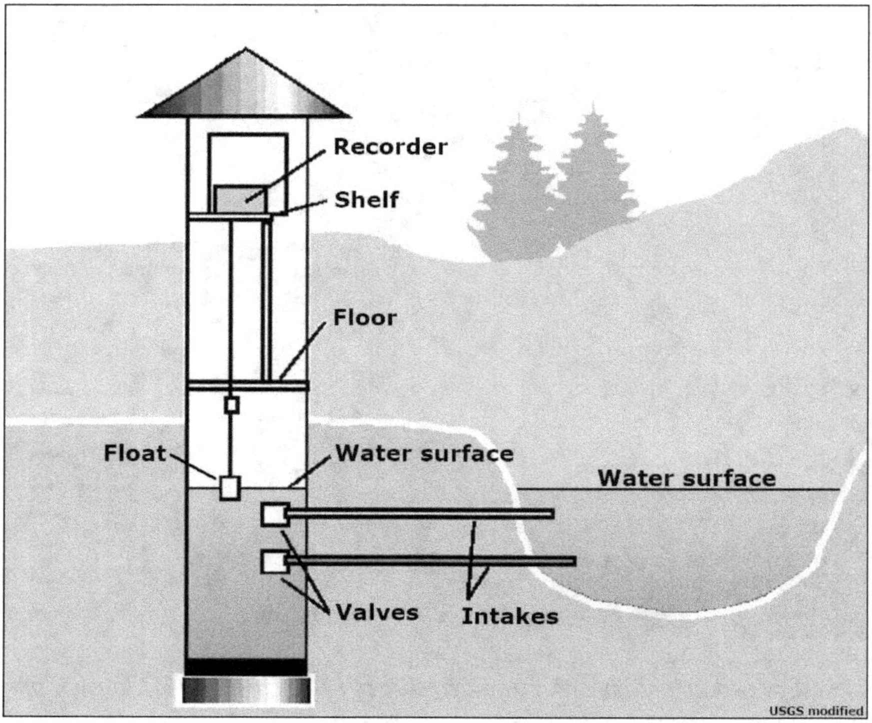

Diagram of a "float" style river gauge. The water surface under the gauge house is the same level as the river surface (U.S. Geological Survey)

"datum".[3] The depth of a river varies with deeps and shallows, depending on the river bottom, while stage is an absolute measure that can be correlated to the river's "discharge", or "flow", expressed in cubic feet per second (cfs).

There were back-up methods to measure the stage if the bubble gauge didn't work, or if Souchik wanted to double check. He had a calibrated tape to probe the water level beneath the gauge house, and a wire-weight gauge mounted on the rail of the North Street Bridge could be reeled down to the river surface. Finally, a staff gauge – essentially a big ruler – affixed to the concrete pier that anchored the bridge to the riverbank reported the stage to the tenths of a foot. None of these devices was foolproof, but Souchik couldn't imagine a circumstance in which none of them would work.

This day, though, the bubble gauge worked fine. The display read 4.81 feet above the local zero point. Souchik pulled a blank data card from the pile on the shelf and entered the figure with the pen he kept in his shirt pocket. He noted that despite yesterday's rain the river surface was nearly a foot lower than it had been the day before. The quick downpour had not raised the river level.

Finished with the river measurements, Souchik stooped to check the rain gauge mounted on a metal frame a few feet from the gauge house. The water line in the tube stood at 1.06 inches, all of it from the big storm the day before. There had been rain on six of the ten days in June so far, and the total rainfall was already above the average for the entire month.

On most days, Souchik simply dropped his data card into the mail to Harrisburg. But when there had been more than a half inch of precipitation, like this day, or if the river was much higher than normal, the RFC wanted him to phone his data in as soon after 7:00 a.m. as possible.

Souchik blew into his hands for warmth as he walked the thirty yards to his Civil Defense office in the sub-basement of the courthouse. He would call his data in to Harrisburg, then go to church in plenty of time for choir practice. Olga would be pleased.

The temperature at Wilkes-Barre rose to sixty that afternoon. The polar front that brought frost to Wilkes-Barre, and wrought a flash flood disaster in South Dakota, weakened as it swept south to the Gulf of Mexico. On that day neither Nick Souchik nor anyone else could imagine that its distant echo would, in another week's time, return to the Wyoming Valley with unbelievable fury.

JUNIOR HYDROLOGIST

Saturday, June 10, 1972
Harrisburg, Pennsylvania

Myron "Mike" Gwinner parked his old station wagon in the open lot on City Island, a sixty-acre expanse in the middle of the Susquehanna. Ballfields and a little amusement park dominated the north end of the island, while the City Marina occupied the south. Gwinner was one of the junior hydrologists at the River Forecast Center, and didn't rank high enough for a parking spot at the Federal Building.

Wind gusting down the river cut through Gwinner's jacket as he walked from the island across the Walnut Street Bridge – the locals called it "Old Shaky" – then up a gradual hill to the Federal Building four blocks away. As at Wilkes-Barre a hundred miles to the north, yesterday's storms ushered in an unprecedented June chill.

Harrisburg, Pennsylvania. The Susquehanna River and City Island in foreground, Walnut Street Bridge at left, State Capitol at far left, and the Federal Building a black cube at left-center (Design Pics Inc./Alamy)

Like much like the rest of the nation, Harrisburg in 1972 was a city divided: divided over the war in Vietnam, between young and old, rich and poor, Republican and Democrat, black and white. Less than three years before, anger in the African-American community erupted after the assassination of Martin Luther King, Jr. Three people were killed in the riots, and the National Guard was called in to enforce order.

Despite the divisions, there were signs of hope. Government employment at the state capital provided a floor for the city's economy. Urban renewal projects were underway or planned for the near future. The governor recently moved into the new executive mansion near the river. And, the neo-classical capitol complex across the street from the Federal Building stood as an oasis of grandeur amid the surrounding urban decay.

The RFC's first task every day was to collect river stage and weather reports from the scores of stations in its watersheds. Most of the data came from volunteer observers, with the remainder from dam operators, power plants, and a few official Weather Service stations. There were too many reports for the RFC to handle directly, so receipt of daily data was distributed among three "River District Offices": stations in New York State reported to the Weather Service Office in Binghamton; stations in New Jersey reported to Trenton; and Virginia/Maryland to Washington. The River District Offices relayed their compiled data to the RFC via teletype or radio. Meanwhile, the hydrologists at Harrisburg received data directly from sixty stations in Pennsylvania, including the Susquehanna watershed.

Most of the observers reported by phone, while a few others transmitted their data via two-way radio. In addition, some of the gauges were semi-automated, with a "telemark" system that, when called by phone, answered with a series of beeps indicating the stage of the river. The telemark phone at the gauge was live, and the hydrologist calling might hear passing trains, courthouse bells, even the conversations of pedestrians.

This being a Saturday, only Gwinner and intern Patsy Quigley were on hand to receive the daily observers' reports. Quigley called out to the telemark gauges and handled reports coming in via radio,

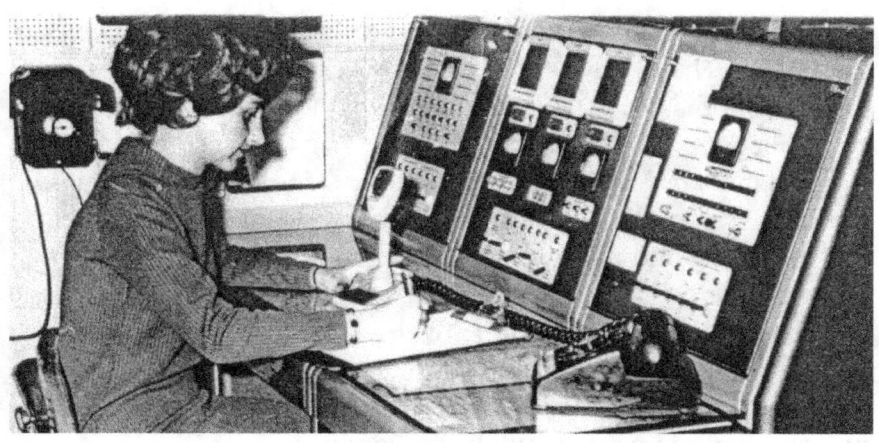

Weather Service intern Patsy Quigley at the radio console of the Middle Atlantic River Forecast Center (NWS, from Shank, Great Floods of Pennsylvania)

while Gwinner received the data phoned in by observers. The RFC had a six-line telephone system, each phone with an old-school rotary dial above a row of buttons that lit up as a call came in on the corresponding line.

Quigley arrived early and started the coffee. She recently graduated from college, and would be assigned to three one-year stints at different Weather Service offices. In the 1970s a professional woman in the Weather Service was still a novelty, and the otherwise all-male staff was trying to adjust.

Gwinner no sooner poured himself a cup and settled at his desk when "Line 1" on his phone lit up. He cradled the receiver to his ear and punched the blinking button. "RFC, Gwinner." The caller was Nick Souchik, at Wilkes-Barre, often the first observer to report. Gwinner never met Souchik in person, but felt as if he knew him from their telephone discussions over the last few years. Gwinner established a rapport with many of the volunteers, even though most days there was time for only a minute or two of conversation. This back-and-forth was more than simple cordiality, but good science as well. The observers were, after all, the RFC's eyes in the field, and had a better sense than anyone for the conditions at their stations.

"The best observer is someone who lives along the stream," Gwinner said later. "They know exactly how far the water is from coming into their home."

Souchik reported his numbers, reading them off the data card he filled out a few minutes earlier. "River stage 4.81 feet and falling. Precip 1.06 inches." Then, extending the conversation, he remarked that it seemed odd the river would be falling after so much rain.

Gwinner responded that most of yesterday's rainfall had been absorbed by the soil, and didn't flow into the river. But, like a kitchen sponge, the soil could hold only so much. Without saying it aloud, he considered that with so much rain lately the soil-sponge was nearly full, and that more rain could bring flooding.

The "Line 2" button on Gwinner's phone lit up in mid-sentence. Gwinner, congenial and garrulous, almost never finished a conversation before the next observer called in. After jotting the Wilkes-Barre data onto the daily summary form, he punched the blinking button on his phone. "RFC, Gwinner."

FLASH FLOOD

Monday, June 12, 1972
Harrisburg, Pennsylvania

The headline of the Harrisburg Patriot-News cried *South Dakota Flood Toll Reaches 208*[1]. Details about the disaster at Rapid City were coming in.[2] On the night of June 9 thunderstorms over the Black Hills brought torrential rain to the headwaters of Rapid, Battle, and Box Elder Creeks, swelling those streams to levels never seen before. Hundreds of homes, automobiles and campsites were demolished as the flash flood swept through the darkness. So far more than two hundred deaths had been counted, many of them people trapped in their cars as flood waters roared through the city. Exhausted rescuers continued to work around the clock, frantically looking for survivors amid the debris. It was one of the deadliest natural disasters in U.S. history.

To O.D. White, the Rapid City flood was more than an abstract tragedy. The Weather Service – especially the River Forecast Centers – was in business to predict this kind of event, and to warn people who might be in harm's way. To be sure, White's Middle Atlantic RFC had no responsibility in South Dakota. Yet his feeling that his agency had somehow failed could not be denied. He clamped his lips on the stem of his pipe as he looked over the preliminary reports from headquarters. White would do everything in his power to make sure nothing like this happened on his watch.

What had gone wrong at Rapid City? White was not quick to place blame. He knew that nature could do the unexpected, overwhelming plans and defenses set in place by mere humans. The Black Hills disaster would prove to be the result of a freak storm in the headwaters and too many people in the flood plain below. Still, there were lessons to be learned, lessons that would better prepare the River Forecast Service for the next storm.

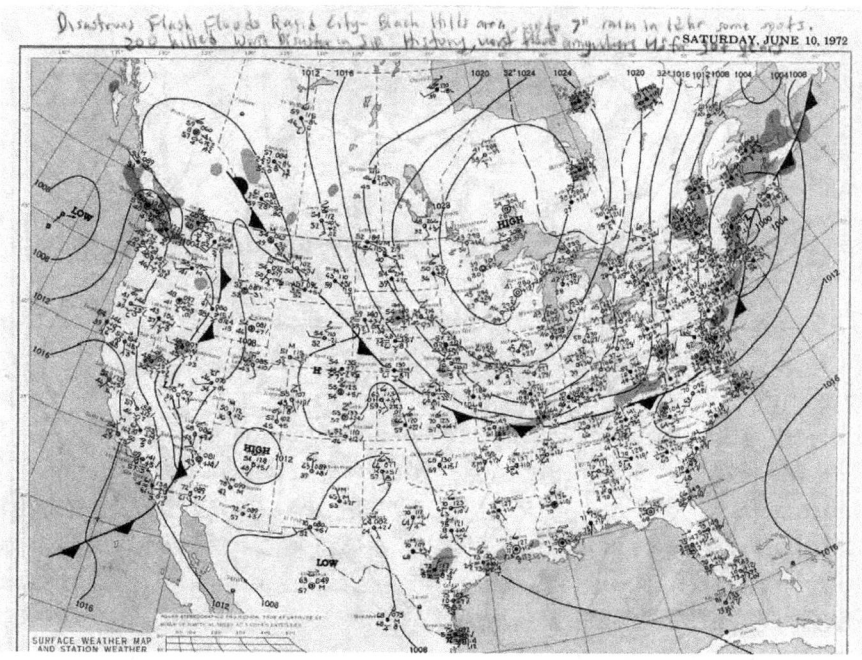

Daily surface weather map, June 10, 1972, with the extraordinary notation at the top: "Disastrous flash floods Rapid City - Black Hills area, up to 7" in 12 hours. Some est 200 killed, worst disaster in S.D. history, worst flood anywhere in U.S. for 30+ years." The powerful polar front that brought floods to South Dakota and record cold to the Middle Atlantic states would days later kick-start formation of Hurricane Agnes (NWS)

White continued to read the reports with keen interest. On the morning of Friday, June 9, the Weather Service at Rapid City predicted isolated thunderstorms over the region. An enormous mass of cold air was surging southward along a 2000-mile-wide front, displacing warm, humid air ahead of it. The western flank of the front approached the Dakotas, while the eastern boundary brought frost to Wilkes-Barre and record low temperatures throughout the Middle-Atlantic states.

The storms built quickly over the Black Hills, then, unexpectedly, stayed for hours. Beginning in the early evening ten inches of rain fell over the headwaters of Rapid Creek and other streams, *four times* the amount of rain that could be expected once in a hundred years.[3] Runoff over the shallow soil and bare rocks in the Black Hills quickly swelled Rapid Creek from its normal discharge of 165 cubic feet per second to 50,000 cfs, far exceeding the previous record flood.

White wondered if the Weather Service received timely information about rainfall in the hills. That, it turns out, was the problem. In 1972 there were two ways to know how much rain was falling. First, rain gauges monitored by the Weather Service or volunteer observers could report rainfall in real time. Yet in the sparsely populated Black Hills that day there were no observers at all.[4] A few concerned citizens phoned in reports of water over the roads, but there was no official measurement of the rain as it was falling.

White had the same problem in parts of his region. While he had more than two hundred observers in the five-state area covered by his office, there were very few up in New York's sparsely-populated Southern Tier. No one wanted the job for the $3.60 per month stipend the Weather Service offered. The observer at Corning, New York, a key station on the Chemung River, recently retired, and White hadn't yet found a replacement.

The second way to measure rainfall was by radar. During World War II it was discovered that the new radars looking for enemy aircraft could also "see" precipitation. By 1972 a network of "Weather Service Radar - 1957s" had been installed at fifty locations around the country. Unlike the Doppler radars with their multi-colored displays seen on television years later, the WSR-57 required an operator to sit and watch the scope go around and around. It took training and experience for an operator to translate what he saw on the scope into a quantity of precipitation. And, while the rate of rainfall could be estimated within a hundred-mile radius, beyond that distance the best the observer could say was whether or not it was raining at all.[5]

The radar coverage at Rapid City was shared with military radar at nearby Ellsworth Air Force Base. Of all days, the radar operators at Ellsworth were having trouble with the equipment. The scope functioned only intermittently that crucial afternoon.[6] Ellsworth got

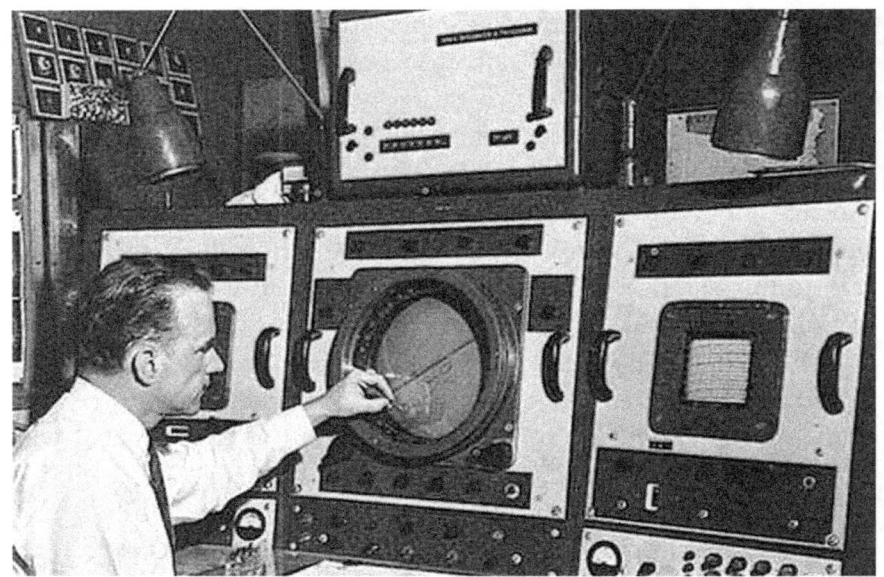

WSR-57 radar scope. A skilled operator could estimate precipitation within a 100-mile radius (NWS)

reports off to the Weather Service only twice, delayed by hours, indicating the heavy rain that was falling over the hills.

White reflected on radar coverage in his Mid Atlantic region. There were WSR-57s at Buffalo, Pittsburgh, New York City, Atlantic City, and Patuxent River Naval Air Station. These installations covered most of his region. But there were some areas the radars did not reach, or could not show how much rain was falling, and he worried about these gaps. The most glaring was, again, over the rugged hills of southwestern New York State, headwaters of the Allegheny, Genesee, Chemung and West Branch of the Susquehanna Rivers. White's hydrologists would be hard-pressed to predict flooding in the headwaters without good rainfall data from the hills of New York's Southern Tier.

The Hydrologist-in-Charge at the Missouri Basin RFC, headquartered in Kansas City, issued a flash flood warning for the streams of the Black Hills at 7:18 p.m. Friday evening. Local police and National Guard mobilized to warn residents of the impending disaster. Radio and television stations broadcast constant alerts, and emergency sirens blared throughout the city. These efforts saved lives, but the worst of the flood came at midnight, when few people were awake to hear, or heed, the final dire warnings.

The terror at Rapid City was wrought by ten inches of rain over a watershed of only 460 square miles. As White looked from his office window to the Susquehanna he thought of the disaster that would ensue if that much rain fell over the 27,500 square miles of its watershed. He knew there had never been so much rain over such a huge area, but the incredible catastrophe at Rapid City made him all too aware that it *could* happen.

* * *

Thursday, June 15, 1972
Yucatan, Mexico

The great temples of Chichen Itza stood on Mexico's Yucatan Peninsula for centuries. Abandoned by its builders 800 years ago, the ancient city had recently come back to life. Now, it was international tourists, not Maya shamans, who walked among the towering pyramids.

This day the magic of the Mayas was drenched by a pounding rain. Only a handful of tourists ventured about the grounds, and the few who tried were fast soaked to the skin as gusty winds turned their umbrellas inside-out. Glowering grey-green clouds scudded by above them, punctuated by flashes of lightening and crashing thunder.

Mexico's *Servicio Meteorologico Nacional* predicted continued heavy rain on the Yucatan, as its data indicated a center of low atmospheric pressure forming almost directly over the ancient ruins. The rotation of the Earth imparted a spin to the winds as they rushed toward the center, and the entire system began a counter-clockwise rotation. The storm had become, in the parlance of meteorologists, a "tropical depression."

Meteorologists would later reconstruct the events leading to formation of this tropical cyclone over the Yucatan. The cluster of thunderstorms induced by El Nino off the coast of Ecuador drifted north into the Caribbean Sea. At the same time, the powerful polar front that days earlier brought devastating floods to Rapid city and frost to Wilkes-Barre spread south over the Gulf of Mexico, pushing its own towering storms ahead of it. Convergence of these two systems triggered development of the unified storm that would become Hurricane Agnes.[7]

The tourists were soaked by the rain as they boarded the little bus that would take them back to the Mayaland Hotel. There they would change into dry clothes, then retire to the bar to relax with some cold *cervezas*, regaling each other about the wonders they had seen that day.

Convergence of a storm cluster from the Pacific and the polar front sweeping down from the United States led to formation of a tropical depression over the Maya ruins at Chichen Izta, Mexico (U.S. Army Corps of Engineers, Tropical Storm Agnes!)

SALVANOS!

Friday, June 16, 1972
Miami, Florida

In 1918 a six-year-old Robert Simpson clung to his father's back as they swam through the storm surge of a monster hurricane in Corpus Christi. That traumatic experience fueled Simpson's life-long quest to understand these dangerous storms,[1] and it is no exaggeration to call him the father of modern hurricane science. He remains famous to this day as co-author of the familiar Saffir-Simpson scale that ranks hurricanes from Category 1 to 5, based on their wind speed and the severity of damage they can cause.

On this afternoon in 1972, Simpson, director of the National Hurricane Center (NHC), presided over a meeting with hurricane specialists Paul Hebert and John Hope[2] at NHC's gleaming new headquarters on the campus of the University of Miami. Representatives of the National Environmental Satellite Service, the Air Force, and other branches of the Weather Service were there too. Papers, charts and maps were strewn about the big conference table: data from radar, land and balloon observations, reconnaissance flights, records of historic hurricanes, and the output of computer weather models. All eyes were on the "tropical depression" drifting into the Straits of Yucatan.

The lights dimmed as Simpson turned on the overhead projector. He called on the satellites expert to explain the latest images. There were two satellites with cameras trained on the storm: ATS-3, in geosynchronous orbit 23,000 miles above the equator, and ESSA-9, in polar orbit 900 miles up. ESSA-9 circled the globe twelve times every day, passing over the Gulf of Mexico twice during daylight. Its most recent image showed the storm as a comma-shaped formation of clouds, the classic incipient spiral that can grow into a hurricane, mid-way between the Yucatan Peninsula and the western tip of Cuba.

Robert Simpson,
Director of the National
Hurricane Center
(NOAA)

The clouds spread hundreds of miles to the east of the storm center, reaching all the way to southern Florida. The specialist analyzed the technical details about the image and reported the schedule for the next passes of the satellites. The Satellite Service promised to give this new storm top priority.

The Air Force representative came next. Reconnaissance flights had been investigating the developing tropical depression for two days, and the latest report was just in. Copies were passed around the table, and an enlargement projected on the screen. The air recon included observations of temperature, humidity, barometric pressure, cloud types, wind direction, and wind speed at several locations and altitudes in the storm. All of the data was crucial for entry into the computer models that predicted the storm's path and intensity.

The wind speed reported by the reconnaissance flight was a special interest, as it determined the nomenclature to be applied to the storm. When the winds of a tropical cyclone exceed thirty-nine miles per hour, the system is called a "tropical storm" and is given a name by the National Hurricane Center. The name makes for easy reference by meteorologists, and gives the public, and media, a ready handle on the storm. Only when wind exceeds seventy-four mph can the storm be called a hurricane. The names for tropical storms and hurricanes are selected years in advance, and everyone at the meeting knew what they were.

Until this morning, June 16, the highest winds in this system clocked in at only thirty mph, but the latest air recon reported winds up to forty-five. What had been a cluster of thunderstorms drifting up from the equator strengthened after meeting the polar front descending from North America, began to spin around a center of low pressure over Mexico, and intensified over the warm water of the Straits of Yucatan. A tropical storm was born, the first of 1972. "And so," Simpson said as the meeting broke up, "we have Tropical Storm Agnes. We're going to have a busy few days. Let's get to work".

* * *

Friday, June 16, 1972
Mantua, Cuba

The landscape of Pinar del Rio, the westernmost province of Cuba, is so spectacular it has been designated a World Heritage Site. Huge limestone hummocks, small mountains known as *mogotes*, loom up from a green plain dotted with colorful villages. Lush forests teeming with unusual wildlife shroud the hills and steep valleys between. Residents call it *El Jardin Natural de Cuba*.[3]

A score of rivers braid down from the mountains, north and south to the coasts. None of them measures more than thirty miles long, and none are notably wide. But heavy rain in the hills causes the rivers to rise quickly and high, carrying with them a load of silt that fertilizes the plains below. This periodic deposition replenishes soil exhausted by the region's most important crop. Any cigar aficionado will affirm that the very best tobacco in the world is grown in Pinar del Rio.

Tobacco fields among the mogotes in Pinar del Rio, Cuba (Jane
Sweeney/RobertHarding)

Claudia Delgado sat at the table in the outer room of her small
home, listening to the rain pounding relentlessly on the tin roof.
What started as breezy showers the night before was now a wind-
driven deluge. She fidgeted with her rosary beads and glanced at
the two portraits on the stucco wall, one of the Virgin Mary and
the other of Fidel Castro. Luis should have been home by now. Her
husband walked into town this morning to certify transfer of their
tobacco crop from the regional co-op to the state-run drying house.
Luis and Claudia were permitted to own their small farm, but every
ounce of production had to be accounted for. She expected him home
an hour ago, but still there was no sign of him.

The Delgado home stood on a slope that tilted gradually down
to Rio Mantua. No one built houses right next to the river, because
it flooded too often. The Delgado farm was overdue for a flood,
and probably couldn't produce another good tobacco crop until one
came. Some of the fields along the river where last year they grew
tobacco were this year in grass, serving as pasture for their small
herd of Brahman cattle – ten cows and six spring calves.

Luis burst through the door, water dripping from his poncho. He had been soaked during the two-mile walk from the village. As poor farmers, the Delgados could not afford a car. Electricity had not yet come to their ramshackle home, and they had no television or even a radio. "Claudia," he blurted out. "I stopped at the community center." The center was the social hub of Mantua, and there was a television there. "This rain is bad, a hurricane maybe." There was fear in Luis' eyes. Every Cuban knew the danger a hurricane could bring. Just three years before, an incipient Hurricane Camille brought ten inches of rain, forcing 20,000 residents of Pinar del Rio to flee their homes.

"Get the cows!" Claudia jumped up. The small herd was the only thing of value Luis and Claudia owned. She remembered her grandfather's advice: *When the rain comes, don't wait to move your livestock. If the river floods, you could lose your animals, and maybe your own life trying to save them.*[4] As she pulled her poncho over her head and followed her husband out the door, Claudia glanced back at the two portraits, murmuring, "*Salvanos!*" Save us.

Five inches of rain had fallen in the hills above the village, and Rio Mantua flowed brown with silt. As Luis and Claudia ran down the slope from their house they could see that the river was already over its banks and creeping higher. Their cattle huddled in an agitated group on a small rise, now a little island, only twenty yards from the riverbank. It was clear that the overflowing river would soon reach them. Luis and Claudia sloshed through the normally dry swale below the rise, waving and shouting at the frightened cattle, "*Vamonos, vamonos!*"

Luis got a rope around the neck of one of the cows and tugged her away from the river, while Claudia prodded the herd with a stick from behind. The calves lowed plaintively, close to their mothers' sides.

It took only five minutes to get the cattle moving, but in that short time the shallow swale below the rise became rushing whitewater. Luis tugged the lead cow and her calf through the knee-deep torrent. Sensing safety, the rest of the herd followed, but the last calf balked at the gushing water. Frantic, Claudia put her arms around the animal's neck and tried to wrestle him across. The frightened animal bucked and shook his head, knocking Claudia from her feet as she released her grip. She struggled to stand, but couldn't regain her footing. The current was too strong, and she was carried down the swale toward Rio Mantua.

Luis ran alongside the swale to catch up to his wife. He tried to throw her the rope, but it wasn't long enough. The swale was deeper as it neared the river, and the water continued to rise. Luis waded into it, almost to his waist, Claudia just beyond his grasp, until he too slipped on the muddy bottom. Somehow Luis dragged himself to safety, but by then Claudia was out of sight. He ran further beside the swale, then almost a mile along the roaring Rio Mantua. "Claudia, Claudia!" The only sound in return was of the wind-driven rain and the raging river. Claudia Delgado was never seen again, and her body was never recovered. Agnes claimed its first victim.

Almost sixteen inches of rain fell in Pinar del Rio by the time Agnes was finished with Cuba. Rio Mantua, Rio Cuyaguateje, and other rivers rose to record floods.[5] The farming communities of Mantua and Guane were hit hard as the spreading waters inundated 45,000 acres of farmland, washed away scores of bridges and roads, and flooded hundreds of homes. More than 35,000 residents evacuated low-lying areas ahead of the rushing waters. Many of the tobacco barns, newly filled with the year's crop, were destroyed, wiping away the main source of income for most of the population. It remains uncertain how many lives were lost; at the time, Radio Havana said seven died,[6] while Cuba's National Institute of Water Resources later reported sixteen dead and twenty-four missing.[7]

HURRICANE HUNTERS

Saturday, June 17, 1972
Ramey Air Force Base, Puerto Rico

Capt. Frank Sanderson powered up the four big turbo-prop engines of his WC-130 aircraft on the tarmac of Ramey Air Force Base, home of the 53rd Weather Reconnaissance Squadron[1]. Weather officer George "Hank" Henry, a co-pilot, navigator, and payload specialist were on board. Their objective was Tropical Storm Agnes, churning in the Gulf of Mexico west of Cuba.

Lockheed's C-130 "Hercules" cargo plane came into service in 1956, and soon gained a reputation as a go-anywhere, do-anything aircraft. Versatile, durable, and easy to fly, it found service as a bomber, parachute trooper, tanker, medical evac, fire-fighter, and VIP transport. Almost seventy years later, with state-of-the-art upgrades, the C-130 remains a workhorse of military air cargo service around the world. In 1963 the Air Force ordered a few 130s specially modified for weather reconnaissance, giving it a "W" in front of C-130.

Sanderson pitched the plane up sharply as it lifted off the runway. Without heavy cargo the WC-130 was light and climbed quickly. It would be a long flight – 1,200 miles over four hours – to their rendezvous with Agnes.

The flight path took them along the southern edge of a huge system of disturbed weather north of Puerto Rico, a thousand miles to the east of Agnes. This storm spawned gale-force winds and heavy seas, interfering with the biennial Newport-to-Bermuda yacht race but not causing any other harm. Indeed, this eastern storm would gain its own cyclonic spin, and would this very day be elevated to "tropical depression". Saturated air from this new storm in the Atlantic was spun westward into Agnes,[2] hugely expanding its diameter and increasing the moisture available to fall as rain.

Satellite image June 17, 1972, showing the incipient Hurricane Agnes at left and a secondary tropical depression to its east (NOAA)

Sanderson skirted the north shore of Cuba, since he was not permitted to fly over the island itself. Just after 7 a.m., over the southern Gulf of Mexico, the plane flew into an outer band of Tropical Storm Agnes. Hit by a sudden gust almost head on, the big plane rose suddenly, then dropped as the gust passed. If uninitiated passengers had been aboard they would have reached for their airsick bags, but this crew had flown into storms many times before, knew what to expect, and seemed immune to the turbulence. It would be a bumpy ride for the next two hours, but the crew, and their aircraft, had seen worse.

On cue from Weather Officer Henry, the payload specialist fed multiple drop-sondes through the launch chute and into the storm. A drop-sonde is a cylinder two feet long and four inches in diameter, packed with sensors that measure weather data as it parachutes to the sea. Data is radioed from the sonde to the hurricane hunters above, measuring atmospheric conditions that will be imported to the models forecasting the storm's path and development. Capt. Henry dispatched twenty-five drop-sondes into Agnes on this flight, and all but one of them worked perfectly.

At 8 a.m., Capt. Sanderson piloted through the towering bank of thunderstorms surrounding the very center of the storm, the "eye wall", and broke into the relative calm of the eye. The eye was unusually wide, as is typical of weak hurricanes. It took the flight almost six minutes to cross it before re-entering the wall on the opposite side. This was their last pass through Agnes before they headed back to Ramey.

As the flight emerged from the turbulence of the storm and into calm skies beyond, weather officer Henry radioed his report back to Ramey. The eye was a well-formed circle 30 miles wide. Its central pressure at sea level was 978 millibars, the lowest recorded in Agnes thus far. Crucially, Henry clocked sustained winds in the eye wall at seventy-five miles per hour. Now exceeding, barely, the threshold of seventy-four mph, the storm was a hurricane – *Hurricane* Agnes.

The communications officer at Ramey tele-typed Henry's data to the National Hurricane Center in Miami, where Director Bob Simpson issued a press advisory.

AGNES REACHES HURRICANE STRENGTH
THREATENS WESTERN CUBA AND LOWER FLORIDA KEYS

The advisory continued: Agnes was expected to grow stronger, and was on a path toward the Florida panhandle. Residents of the Keys and the Florida peninsula were urged stay tuned for further updates.

Agnes became the only tropical storm or hurricane to hit the U.S. mainland in all of 1972. In late August, Tropical Storm Carrie brushed the New England coast, and in mid-September Hurricane Dawn, described as a "trivial hurricane with trivial consequences",[3] meandered past the Outer Banks of North Carolina. Neither storm caused any harm. It was an extraordinarily sparse year for Atlantic storms and hurricanes. Scientists did not yet realize that an El Nino in the Pacific, as there was in 1972, often correlates with diminished storm development in Atlantic Basin.[4] The cause-and-effect of this correlation is still not fully understood.

* * *

Saturday, June 17
Rapid City, South Dakota

The stricken residents of Rapid City were just beginning to recover from the devastating flash flood a week before. They were still sorting through the wreckage of their homes, still searching for survivors, still burying their dead. But now the Weather Service warned of another flash flood. Thunderstorms that came with a new cold front rolled over the Black Hills. Forecasters predicted a four-foot wall of water could rush down Rapid Creek and again inundate areas still buried in mud and debris. Sirens wailed throughout the city. Mayor Barnett declared martial law, and recalled 2,500 National Guard troops who had been packing to leave.[5] Beyond anxious, the dazed residents of Rapid City were scared witless, and ran to high ground or scrambled to their rooftops.

This time the thunderstorms moved quickly past. There was no wall of water, and only minor damage. One man was killed when he drove his car off a flooded road. With a weary sigh, the people of Rapid City got back to the dirty and tragic work of recovery.

The cold front that scared South Dakota on June 17 marched on towards the southeast, displacing the humid air ahead of it. In a few days it would bring its thunderstorms and rains to the Middle-Atlantic states, where it would dance an atmospheric tango of death with the storm named Agnes.

GENERATION GAP

Bill Shock looked in the mirror and ran his fingers through his growing mustache, hoping the girls on Public Square would think him older than his nineteen years. He was glad it was a weekend. Shock landed a job at the big RCA plant in Mountaintop after graduating from Wilkes-Barre's Meyers High School two years before, but cleaning tools and parts for electronic devices was a drag. He was looking forward to the party on the Square.

Shock lived with his aunt and uncle in their double-block home on Edison Street, staying on in Wilkes-Barre after his parents separated and moved away. When September came he planned to begin college at Penn State's campus in Norristown. Shock wasn't thrilled about going back to school, but his parents insisted, and it would gain him a 2-S deferment from the draft. Now, though, the main thing on his mind was tonight's party downtown. He shook his head to bounce his thatch of wavy dark hair.

Every Saturday night the youth of Wilkes-Barre gathered on Public Square, a grassy, shaded oasis in the urban center of the city. Shock and one of his friends headed that way about 8:00 p.m., stopping at a bar for a six-pack of Stegmeier's. His friend was old enough to buy it, and Shock fancied himself old enough to drink it. They arrived downtown just as the sun was setting.

Already a hundred young people gathered on the square. Most of the boys wore longish hair, and many of them sported a scraggly beard or fu-Manchu mustache. Every one of them wore a tee shirt and blue jeans, some with fashionable holes in the knees. The girls dressed in tie-dyed tees and ragged cut-offs, showing way more leg than was acceptable in conservative Wilkes-Barre. Pink Floyd and

The Rolling Stones throbbed from portable radios as the sun went down. It wasn't an organized party, but there wasn't much else to do in Wilkes-Barre on a Saturday night. The kids were "hanging out", chatting and flirting, mellow as the warm summer breeze. Here and there a whiff of marijuana smoke floated in the air.

This was not the way the decent folks of Wilkes-Barre thought young people should dress and behave. Older residents were hostile to the counter-culture seen in the hairstyles, music and political views of these youths. "Hippies!" passers-by would spat, or, often, "Damn hippies!" Just in the past week a group of college kids, home for the summer, were chased from a downtown burger joint as the owner shouted obscenities.[1]

Shock and his friends downed their Stegmeiers as the party got louder. He was chatting up a young woman with long dark hair under one of the tall elms when they heard a ruckus at the far side of the square. There was shouting, then moments later, a police siren.

"We'd better go," the girl said, then quickly added, "What's your phone?" She fished a pen from her purse and wrote Shock's number on her hand. "I'm Ellen. Ellen Rogers. I'll call you."

The following day the Wilkes-Barre Times-Leader featured an editorial condemning *Hooliganism on the Increase Here*, perceived to be rampant in the city.[2] It cited examples of minor vandalism, and urged citizens to "get involved" if they saw suspicious individuals or groups of young people. Any gathering of youth, the editorial implied, should be regarded as "suspicious".

Truth be told, there were few "hippies" in Wilkes-Barre. Most of these kids came from hard-working blue-collar families, many with conservative social values. Their parents expected them to go to church or Synagogue every week, as they had done all their lives. And while they might not support the Vietnam War politically, they did support their friends, family, and neighbors in uniform. Shock's cousin had been drafted and just finished a two-year stint in the Army, including a year in 'Nam, and the entire family was enormously proud of him. Yes, the local kids affected the music and styles of their peers, but the sons and daughters of the Wyoming

Valley were far more likely to stay close to home than join the communes of Greenwich Village or Haight-Ashbury.

* * *

Monday, June 19, 1972
Wilkes-Barre, Pennsylvania

It was a glorious day in the Wyoming Valley for the first time in a week. Frank Townend, wearing an American flag pin the lapel of his dark suit, stepped from his law office across the street from Public Square. He was on his way to a luncheon with Red Cross volunteers at the Sterling Hotel, two blocks away. He wouldn't stay long, just put in an appearance, shake a few hands, and say hello to some friends. Today, too, would be his first chance to meet Eliot Knauer, the state's new, very young, Deputy Secretary of Public Welfare, keynote speaker at the luncheon.

Townend was well known in Wilkes-Barre. A decorated hero of World War II, he was now a prominent lawyer, on the board of Nesbitt Hospital, commander of the National Guard, and chairman of Luzerne County's Council on Civil Defense. He unsuccessfully ran for election as judge a few years prior; though competent and credentialed, some voters found him aloof and inscrutable. [3] Now, though retired from the Army, everyone still called him "General".

As its name implied, Townend's Council on Civil Defense primarily focused on Cold-War military issues, including protection against and recovery from a potential nuclear attack. Most of its paid and volunteer personnel had connections with the National Guard or other armed services, and its operations had a distinct military air to them. Frank Townend, as commander of the local National Guard, was a logical choice as its chairman. The position was voluntary; Townend's role was oversight, while day-to-day operations were managed by paid director Nick Souchik.

General Frank Townend, with the Luzerne County Courthouse in the background (courtesy of Wilkes-Barre Times-Leader)

Seven stories high, the Sterling was once the grand dame among the proud buildings that lined River Street. Built in 1897, a flat-roofed palazzo with a white limestone facade, the Sterling offered the most luxurious accommodations in the Wyoming Valley. The hotel was at the heart of what would later be designated the River Street Historic District. Anchored at the north by the county courthouse, and to the south by Wilkes College, the district had been home to the coal barons during the city's economic heyday eighty years before. Here, overlooking the river, the mansions of the wealthy alternated with magnificent churches, Beaux Arts office buildings, and the towering minarets of Irem Temple, a Masonic lodge built to resemble a Moorish mosque.

The Sterling Hotel boasted a two-tiered lobby, grand marble staircases and a sky-lit atrium. It stood at the historic Market Street Bridge that spanned the Susquehanna. Guests with west-facing rooms enjoyed magnificent views of the river. Yet, like much of Wilkes-Barre, the Sterling showed its age. Townend didn't have to look too closely to see that the once-rich woolen carpets were frayed, and the many layers of enamel around the ballroom faded and chipped. He knew the building's antiquated plumbing and electrical systems weren't up to code, and the owners couldn't afford to repair them. Nevertheless, the grand old Sterling remained Wilkes-Barre's premier location for meetings and conventions. And, Townend's wife Lenchen owned the Wide Awake Bookshop off the hotel's lobby.

Townend nodded to several business acquaintances as he passed the hotel cafe on his way to the ballroom. A clatter of plates being cleared from the tables and the buzz of conversation greeted him as he entered. The luncheon was a "thank you" for the valley's Red Cross volunteers, and about fifty of them, mostly women, were present. The volunteers being honored stood at the front lines of disaster relief. They, or their forbears, proved themselves in the great flood of 1936 and many lesser occasions, giving aid to the victims of natural and man-made catastrophes locally and nationwide. Only a week before, the Luzerne County chapter coordinated donations of clothing and cash for the victims of the flood disaster in South Dakota.

The keynote speaker, deputy welfare director Eliot Knauer, stood at the podium to speak about his agency's plans for disaster response. Looking over the audience, he smiled as he borrowed a line from a popular TV ad for margarine. "Ladies and Gentlemen, it's not nice to fool Mother Nature!"[4]

Townend turned to look out the tall ballroom window. Across the street he saw the green-painted flood wall, and beyond that the placid Susquehanna River. Mother Nature indeed, right here in Wilkes-Barre.

THE DIRTY QUARTER

Sunday, June 18, 1972
Miami, Florida

National Hurricane Center Director Bob Simpson's secretary buzzed in to say a reporter from the Associated Press was on the line for an interview about Hurricane Agnes. Simpson, working the weekend to monitor the new hurricane, was eager to oblige. Once a storm became a hurricane it grabbed everyone's attention, and public education was his passion.

The reporter led off by asking if it was unusual for hurricanes to form so early in the year. Simpson took it from there. "Actually, there have been thirty-five tropical storms in June over the last eighty-five years, since we've been counting. Ninety percent of them became hurricanes."

"Ninety percent? That would be about thirty hurricanes. I don't remember any of them."

Simpson continued. "That's understandable. It's rare for June hurricanes to be very strong". Then he back-pedaled, perhaps recalling 1957's Hurricane Audrey. "I have to add that statistics can be fickle. Any hurricane can be deadly."[1]

Agnes had already packed a deadly punch in Cuba, and Simpson's cautionary words would soon be realized closer to home.

* * *

Ten miles east of boisterous Key West, Big Coppitt Key was a quiet haven for retirees. By midnight Saturday most of the residents were asleep in their modest bungalows. They knew that Hurricane Agnes churned in the Gulf three hundred miles to the west; the Weather Service posted a "hurricane watch" for the lower Keys, but predicted the center of the storm would bypass them on its northward path. There seemed little reason to be concerned.

The forward right-hand quadrant of a tropical cyclone is known as the "dirty quarter". It is here where the winds are the strongest and storm surge the highest. And often, in the outer rain bands, low-level wind shear – that is, winds blowing in different directions at different altitudes – can spawn tornadoes.

One of those outer rain bands in the dirty quarter, hundreds of miles east of Agnes' eye, passed over Big Coppitt Key just after 2:00 a.m. Sunday. A dark funnel, hidden in the night and the driving rain, descended from the low clouds to rip a diagonal path northward between Second and Third streets. There was no warning. Only Charlie Haas, arriving home from a visit to Miami, saw the tornado hit. "My car stalled in the rain", Haas, a World War II Air Corps veteran, recalled, "when I heard what seemed like a flight of B-17s. All of a sudden, it hit the power lines. There was a big explosion that lit up the whole sky!"[2]

A few blocks away, Phyllis Loren was sleeping when her mobile home tumbled in the wind. Pinned beneath a dresser, she managed to free herself. When she stumbled out she found four neighbors laying on the ground, and another, injured, dangling from a tree.[3]

First responders from Key West found chaos when they arrived at Big Coppitt. Fifty-two homes, including forty-seven trailers, were severely damaged or destroyed.[4] Sheriff Ralph White reported that the mobile homes disintegrated in the vortex of the tornado; some of the homes, their occupants still inside, were hurled into the limbs of the native gumbo limbo trees.[5] More than forty people were injured, some of them badly. "It's unbelievable that anyone came away alive," Sheriff White told reporters.[6]

The Big Coppitt tornado was later determined to be an F-2 on the Fujita scale, with wind speeds between 113 - 157 mph.[7] It remains the strongest tornado ever to hit the Florida Keys.[8]

* * *

An hour later, another F-2 tornado struck Key West itself, skipping along the northern edge of the island, ripping the roofs from several homes and businesses. Fifty people were hurt, though none seriously.[9]

Things got worse for Florida, much worse, later in the day. At historic Fort Denaud, a sleepy village thirty miles east of Fort Myers, retired couple Emil and Evie LeClair relaxed in the living room of their small bungalow. Evie's afternoon soap opera just ended at 4:00 p.m., when WINK-TV, out of Fort Myers, broke in with a news bulletin. The reporter confirmed that a tornado spun out of Hurricane Agnes touched down in the Keys the night before, ripping through homes and injuring dozens of people. The broadcast showed dramatic film of shredded mobile homes, and an interview with a dazed survivor. Then the National Hurricane Center's advisory: the center of the storm would remain well offshore today and hit the Panhandle tomorrow. For the first time, the NHC warned *the threat of a few tornadoes will continue to the east of the storm track.*

The first of the storm's outer rain bands had passed over Fort Denaud two hours before, with gusty winds and a half hour of driving rain. Low, angry clouds continued to race across the sky even after the squall abated. The weatherman on WINK-TV predicted showers on and off the rest of the day as Agnes passed to the west. This was before such a thing as live dopplar radar, and there was no way to know that the second of Agnes' outer bands was about to sweep over southern Florida.

At 4:10, the sky darkened as a blast of rain pelted the LeClairs' house, followed by a flash of lightning, very close, and a boom of thunder. Moments later, said Evie, "it sounded like a freight train was coming."

Emil ran outside and froze in his tracks as an F-2 tornado dropped from the sky. It clipped the top of a grove of pine trees, then ripped into the house across the street, home of Buddy and Vickie Messer and their six-year-old daughter Bonnie. "When it hit, there was a flash and an explosion," Emil told a reporter. "There was nothing left of the trailer." As LeClair ran toward the demolished Messer home he saw little Bonnie running, screaming, down the street. Buddy Messer, blood trickling from a gash on his scalp, stood over Vickie's body amid the wreckage. She was horribly battered, and obviously dead. Buddy looked at LeClair, pleading, "Please help me. Please God, help me."[10] Vickie Messer, thirty-one years old, of Fort Denaud, Florida, was the first person in the United States to die in the fury of Hurricane Agnes.

Seven hours later, at Okeechobee, fifty miles northeast of Fort Denaud, yet another F-2 tornado touched down to scrape a broken path hundreds of yards wide and ten miles long as it hopped from one mobile home park to another. Witnesses said they would never forget the roar caused by the twister. "It was like jet engines splitting our eardrums," recalled Debbie Broome. "We ran out and saw the funnel. It was so black, and somehow illuminated by an eerie blue light."[11] The Treasure Isle, Four Acres, Taylor Creek, and Pine Ridge mobile home parks around Okeechobee were hit hard. Six more people died, and hundreds left homeless.[12]

The tornado outbreak reached the east coast of Florida and northwards into Georgia. As many as thirty twisters[13] were counted, beginning from the Keys in the south, then north along both Florida coasts. Tornadoes associated with tropical cyclones are typically weak, small, and brief, but several of the funnels in Agnes were strong F-2 and F-3 maelstroms. On the "Space Coast" near Cape Canaveral, an F-3 tornado touched down at the Merritt Island Airport, tossed forty-four aircraft as if they were toys, and demolished two hangars and an apartment building.[14] Half an hour later, another F-3 struck the town of Cape Canaveral to wreck the Coast Guard station and thirty-two homes, leaving twenty-three people injured.[15] At North Palm Beach, another F-2 lifted a sailboat from its moorings and dumped it on the roof of a nearby building. Researchers later determined that "Agnes was the largest and deadliest tropical cyclone tornado outbreak in Florida history, and the third deadliest in U.S. history."[16]

* * *

Sunday, June 18, 1972
Washington, DC

Until now Agnes was a Florida storm, as reflected in the headlines of local newspapers:

Angry Agnes Leaves Trail of Death – Fort Pierce News Tribune
Twisters Eye Gulf Counties – Tampa Tribune
Agnes Roars Up Gulf – St. Petersburg Times
Agnes Careens into Panhandle – Naples Daily News

The hurricane did not yet draw much attention outside of Florida. The Washington Post allowed only a brief paragraph on an inside page:

Agnes, first major storm of the 1972 season, swirled to hurricane strength yesterday [June 17], and forecasters posted a hurricane watch for the Florida Keys.[17]

That was about it. Agnes was barely a hurricane, and seemed to pose no threat to anyone north of Georgia.

Meanwhile, a storm of a different kind was brewing in the nation's capital. While Agnes languished in the back-page weather column, the front page of The Post reported, in a small headline below the fold:

FIVE HELD IN PLOT TO BUG DEMOCRATS' OFFICE
Five men, one of whom said he is a former employee of the Central Intelligence Agency, were arrested yesterday [June 17] in what authorities described as an elaborate plot to bug the offices of the Democratic National Committee ... in a sixth-floor office at the plush Watergate.[18]

The Watergate break-in occurred the very day Agnes became a hurricane. Watergate and Agnes played leap-frog for head-lines in the national press until the following week. Then, it was all Agnes, and the Post diverted young reporter Carl Bernstein from the Watergate story to cover the floods.[19]

LANDFALL

Monday, June 19, 1972
Miami, Florida

Data streamed into the National Hurricane Center from the satellites service, from Air Force and Navy reconnaissance flights, from airports and marine terminals around the Gulf, from sounding balloons lofted to measure the atmosphere, and from dozens of observers reporting from their backyards. All of this data would figure into the models repeatedly run by the Hurricane Center to forecast the storm's most likely path and intensity.

Agnes was, so far, a typical June hurricane, formed in the Gulf of Mexico and relatively weak. The *one* thing that made Agnes unusual was the enormous breadth of the storm.[1] Nearly a thousand miles across, Agnes drew moist air not only from the Gulf, but from the Atlantic Ocean on the far side of the Florida peninsula, a stream of moisture reinforced by the newly-formed tropical depression north of the Bahamas.[2] Agnes was not a strong storm, but it held *a lot* of water.

Hurricane Center Director Bob Simpson was the world's leading expert on the ways in which the surrounding environment "steers" a hurricane along its path, and personally wrote the models to forecast a hurricane's course, speed and strength. This day, the models, plus Simpson's long experience with historic hurricanes and deep knowledge of steering currents, predicted that the center of Agnes would continue to move north to make landfall on the Florida panhandle. The exact location of landfall was crucial, as that would determine when and where the worst of the winds and storm surge would hit. Forecasting the time and place of landfall more than a few hours beforehand was, and remains, notoriously difficult.

The Hurricane Center's advisory Sunday night reported that air pressure in the eye of the storm had fallen to 978mb, the lowest so far. It warned that Agnes would hit shore somewhere between Panama City and Apalachicola at around noon on Monday, June 19th, carrying sustained winds of ninety mph with gusts up to 120. Storm surges along the Gulf coast were predicted to reach as high as eight feet, sure to flood low-lying areas.[3]

Residents of offshore islands and beaches all along the Panhandle were urged to evacuate. Tens of thousands of tourists, their vacations cut short, crawled inland on US Route 231 after Panama City Mayor Dan Russell told them to flee their beachside hotels.[4] Residents crammed grocery stores for milk and bread, while shopkeepers hammered plywood over their windows. Bob Smith, civil defense coordinator for Florida's Gulf Coast counties, said that if Agnes kept to its predicted path he would order 235,000 residents to move inland.[5] The Red Cross set up shelters to accommodate evacuees.

Agnes picked up forward speed as it neared the coast Monday morning, but its strength diminished. When the center passed over Cape San Blas early in the afternoon its winds were down to sixty mph, and Agnes was demoted to "tropical storm". Gusts at Panama City hit only thirty-five mph, and the big beach resorts didn't suffer heavy damage. The tourists fled, it seemed, for naught.

Forty miles east of Panama City the winds, waves and storm surge hit harder. Flimsy fishing cabins, oyster houses and vacation bungalows crumbled to rubble along Florida's beachfront between Apalachicola and Carabelle. "I gave up counting how many homes were destroyed, and just put in 'most of them' on my report," said a Red Cross volunteer. "They just opened up like eggs. And what was left of the houses was heaved up onto the road, as if the Gulf had swallowed them then spit them back out".[6]

Simpson absolutely nailed landfall. Although the storm was weaker than forecast, Agnes came ashore Monday less than fifteen miles west and only two hours later than he had predicted Sunday night.[7] Even fifty years later, the National Hurricane Center would be hard-pressed to forecast a hurricane landfall as accurately as Bob Simpson did for Agnes.

The subsequent life of Agnes would not be so well forecast. Its track over the Southeast, its re-intensification, its sudden turn inland,

and the catastrophic rains and floods it was to bring were missed almost up to the time they actually occurred.[8]

* * *

Monday, June 19, 1972
Washington, DC

The Washington Post reported the Agnes story on an inside page: *Hurricane's Tornadoes Hit Florida.*

The lead story of the day, above the fold, was headlined *GOP Security Aid Among 5 Arrested in Bugging Affair.* Former CIA agent James McCord was linked to the Watergate break-in. Reporters Bob Woodward and Carl Bernstein got the byline, their first in covering the affair. Another inside story described *How NASA's Space Shuttle will Work.*

The Post's daily weather column predicted rain Monday and Tuesday in the Washington area, with clear skies Wednesday Thursday. Agnes was not yet in the picture.

* * *

Tuesday, June 20, 1972
Miami, Florida

Agnes sputtered over Georgia Monday evening, deprived of energy from the warm waters of the Gulf. The storm weakened to a mere "tropical depression" with winds at its core below thirty-five mph.

Hurricane Specialist John Hope issued the Hurricane Center's final bulletin about Agnes at 9 p.m. Monday. He predicted that the remnant storm would follow a path eastward over Georgia and South Carolina, then go out to sea and dissipate in the Atlantic. Small craft along the coast as far north as New Jersey were advised to remain in safe harbor for protection from gale winds and heavy seas. Hope's bulletin concluded with what would prove to be a disastrous understatement: *precautions should be continued against moderate to heavy rains near and east of the path of the storm, with the threat of some flash flooding.*

Forecasting responsibility for Agnes was transferred from the Hurricane Center to regional weather service offices when the remnant storm moved off the Gulf of Mexico and over land. As far as the National Hurricane Center was concerned, Agnes was over.

FLASHBACK: CAMILLE

August 20-21, 1969
Richmond, Virginia

Hurricane Camille is remembered as the strongest hurricane ever to make landfall in the United States,[1] packing gusts up to 230 mph[2] as it roared ashore at Bay Saint Louis, Mississippi in 1969. The storm surge topped twenty-four feet and destroyed every structure in its path. More than a hundred lives were lost in Louisiana and Mississippi before Camille moved inland.[3] Oft remembered is the destruction of the Richelieu Apartments at Pass Christian, Mississippi, where the building and twenty-three residents who stayed for a "hurricane party" were swept away by the storm surge.

Yet Camille had a second act, even more deadly, that is not as well remembered. Downgraded to a tropical depression after making landfall on the Gulf Coast, Camille raced up the Mississippi valley then veered eastward toward the Middle-Atlantic states. The Weather Service predicted two inches of rain as the remnant storm drifted over the Blue Ridge Mountains of Virginia. There, unexpectedly merging with a stalled cold front and enhanced by uplift over the mountains, the storm exploded into a freak rain-making machine.[4]

At 3 a.m. on August 20th an alert resident – not an official observer – in Nelson County, Virginia, phoned the Weather Service Office in Lynchburg to report eight inches of rain since midnight. The creek along his property was overflowing. Lynchburg issued a flash flood warning, but few people were awake to hear it.[5]

Up to *thirty-one inches* of rain deluged the remote and rugged terrain of Virginia's Nelson, Rockfish and Amherst Counties between midnight until 6 a.m., one of the all-time meteorological records in the United States.[6] The Weather Service calculated this to be the most rain the laws of physics would allow.[7] Survivors described it

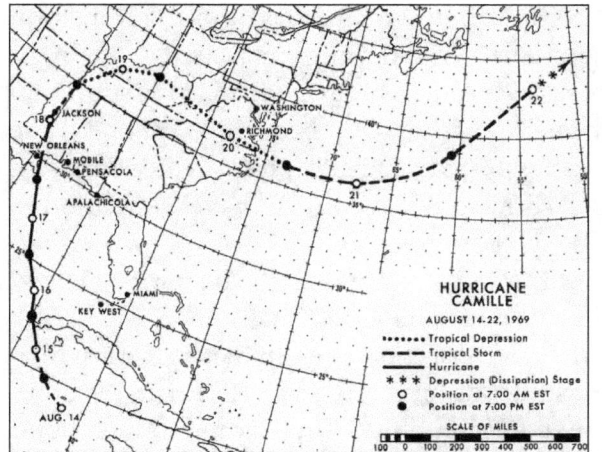

The path of 1969's Hurricane Camille: a Category 5 landfall on the Gulf coast, and devastating floods in the Appalachians (NOAA)

as standing under a waterfall. What were normally trivial creeks became raging, ravenous monsters. Residents went to bed with a forecast of only an inch or two of rain, and there no way to warn them of the looming disaster. At least 153 people were swept to their deaths, more than Camille killed at during its Category 5 landfall in Mississippi. The flood downed all the power and phone lines in Nelson County, and destroyed every bridge. Days went by before rescuers could reach some of the stricken communities. For weeks to come, searchers found drowned and crushed bodies amidst tangles of uprooted trees and debris heaped along swollen streams. More than fifty years later, several victims remain entombed beneath mud and rock, their exact whereabouts forever unknown.[8] It was, and remains, the deadliest natural disaster in Virginia's history.[9]

The Blue Ridge creeks supercharged by the torrential remnant of Camille fed into the James River, which flowed from the mountains a hundred miles to tidewater at Richmond, the capital city of Virginia.

* * *

In 1969, forecasts for the James and other rivers of Virginia were made by hydrologist Michael Mark at the Weather Service Office at Washington.[10] Mark joined the staff only two years prior, and often found himself playing catch-up with inadequate equipment, observer vacancies, a complicated bureaucracy, and the transition to computer modeling of weather and river forecasts.[11]

At 4:30 a.m. on August 20, Mark was awakened at his suburban Maryland home by the persistent ring of his bedside telephone.[12] The caller was Al Kachic, regional hydrologist of the Weather Service's eastern division. "Mike, sorry to wake you. We have a situation in the upper James. Maybe the Potomac too."

Mark tried to clear his mind. "A situation?" The forecast called for only an inch of rain. "What's going on?"

"It's still coming in. There's a report of eight inches of rain up in Nelson County, with flash floods on the Tye and Rockfish. We don't know how bad. Get down to your office. I'm heading to Richmond now, call me there. Good luck."

Mark arrived at his office at 5:30 a.m. to parse the scant data transmitted from Lynchburg and Roanoke. It looked like the Potomac basin would not see serious flooding. But the handful of reports from the headwaters of the James River spoke of incredible rainfall in the mountains of western Virginia. There were sure to be life-threatening floods all the way to down to Richmond.

It would take almost two days for the surge from the headwaters of the James to reach Virginia's capital. Michael Mark, Al Kachic, and their colleagues received enough data from observers upstream to forecast the flood at Richmond well in advance. At 3:00 p.m. on August 20, Mark predicted that the James River would crest at Richmond on Wednesday night, and warned that *all persons should take immediate protective actions.*[13]

City workers piled sandbags to buttress the dikes along Dock Street, an industrial area south of downtown. Teams of Red Cross and city police knocked on doors in the low-lying residential areas and warned residents to evacuate.[14]

Early Thursday, August 22, the James River crested at 28.6 feet at Richmond's City Locks gauge, higher than any time since 1771.[15] Warehouses, factories, stores, and six hundred homes near the river were awash. Yet there were no injuries or lives lost in Richmond, credited by city officials to the Weather Service's advance warning of the flood.[16]

Shortly downstream from Richmond the flood of the James River dissipated into the tidal estuary of Chesapeake Bay. What was left of Camille sped out to sea, briefly regaining "tropical storm" strength before falling apart over the North Atlantic. The storm and flood were gone, but not before bringing the greatest flood at Richmond in two hundred years – a record that would stand only three years more.

AGNES REBORN

Tuesday, June 20, 1972
The Carolinas

Farmers of the Georgia and Carolina coastal plain welcomed the Hurricane Center's prediction for rain east of Agnes's track. Unlike their water-logged neighbors to the north, the southern states had been in a months-long drought. Starved for moisture, tobacco stalks browned and corn shriveled. Some farmers plowed their soybean crop under, lost for the season. Only cotton flourished in the hot, dry weather that hung over the Southeast.[1]

Agnes brought a sudden end to the drought. Five inches of rain fell over the coastal plain, with as much as seven inches along the Georgia coast. The soil got a good soaking, and drying reservoirs filled. The lowland farmers who feared the season would be a bust thanked God for Agnes's passage.[2]

Not as well forecast was the heavier rain west of the storm's center. Though demoted to a "tropical depression", Agnes was still a very wide, very wet, system. Its counter-clockwise winds pulled moisture from the Atlantic and across the lowlands to the foothills of the Appalachians. The sodden air cooled as it pushed up against the mountains. Seven inches of rain fell over a wide area, more than ten inches at Mt. Mitchell, to flood the headwaters of a half dozen rivers that surged down onto the Piedmont. The Yadkin and Pee Dee Rivers swelled well above their flood stages.[3]

The swollen rivers inundated farmland and damaged crops, but flooding in Georgia and the Carolinas wasn't catastrophic. Agnes did, however, claim two more victims in North Carolina: a farmer in Iredell County was swept away when he tried to drive his tractor through a flooded road, despite desperate attempts of bystanders to rescue him.[4] In Surry County, a young man thought it would be fun to take his canoe out on the raging Yadkin River.[5]

He tangled in the branches of a "sweeper" – a tree fallen into the river – and drowned.

* * *

Wednesday, June 21, 1972
Richmond, Virginia

Agnes so far had been a typical early-season tropical cyclone, reaching minimal hurricane strength for less than twenty-four hours before landfall. The Weather Service forecast that the remnant of the storm, now a mere tropical depression, would continue to weaken as it swept over the Carolinas, then move out to sea to dissipate over the cold waters of the Atlantic.

All that changed in the early morning hours of June 21. Unexpectedly invigorated by the still-distant cold front descending from the Great Lakes – the front that scared Rapid City two days before – the remnant of Agnes re-intensified.[6] Pressure at the core of the storm dropped and the winds picked up speed.[7] It's unheard-of for a tropical depression to regain strength over land, without the energy of warm ocean waters to fuel it, but to the astonishment of meteorologists that's exactly what happened.[8] Agnes was a "tropical storm" reborn, its winds again exceeding thirty-nine mph.

With its renewed strength an already-saturated Agnes siphoned yet more water off the ocean into the vast circulation around its center. The winds carried the load of moisture inland and upwards against the flanks of the Appalachian Mountains, dropping prodigious amounts of rain over the Piedmont.

The Weather Service Office at Richmond predicted "showers and locally heavy thunderstorms" for Wednesday, June 21.[9] The showers began as predicted, but grew ever harder as Agnes regained tropical storm strength. By dark the showers increased to a downpour.

Unlike the highly local Nelson County floods from 1969's Hurricane Camille, rains from the reborn Agnes covered the entire range of the Appalachian Mountains and Piedmont regions. Already sodden from three inches of rain only days before, the headwaters of the James, Appomattox and Rappahannock Rivers were deluged with as much as 13.6 inches of additional rain from Agnes.[10] Flash floods on the little tributaries came mostly during the night of June 20 and early morning of June 21. The Weather Service offices in Richmond, Roanoke, Lynchburg and Washington issued bulletins warning of the danger,[11] but, as with Camille, many sleeping residents never heard them. Meteorologists abandoned technical language to capture the essence of the disaster:

> *The real tragedies occurred along the smaller tributaries. Every creek and stream worthy of the name overflowed its banks and wiped clean the adjacent land. Homes and a lifetime accumulation of household goods were quickly swept away*[12]

And lives were lost. Donald Cox, a 61-year old minister, and his wife Mary were driving from Illinois to attend their daughter's wedding in Virginia Beach. They crossed the Appalachians early on Wednesday morning, eager to arrive at their destination that afternoon. Cox came to a halt just after crossing the James River on U.S. 60. Several other cars stopped and blocked the highway. A gaggle of people stood in the road, watching the swelling flood at the confluence of little Bent Creek and the James. The rising water of the creek swamped over the highway along a normally dry slough, and it was impossible to know how deep it was. They could have back-tracked to find another crossing further south, but Cox was determined to be on time for the rehearsal dinner that evening. He drove forward into the water sloshing over the road, despite the warning of other motorists. Moments later, Cox felt the car lifted and pushed sideways by the flood. He threw the transmission into reverse to back out, but the wheels no longer touched the ground. To the horror of the onlookers the car pitched sideways, then rolled off the embankment into the opaque flood water before coming to rest, upside-down, against a tree. Three days later the body of Mary Cox was found ensnared in a hammock at a nearby farm. The car, battered by floating trees and other debris, was barely recognizable when it was towed from the mud, the body of Donald Cox still inside.[13]

Seventeen people were killed by Agnes in Virginia, many of them, like Donald and Mary Cox, swept to their deaths while trying to drive through flooded roadways. The slogan *turn around, don't drown* was not yet in vogue, but as urgently applicable then as it is now.

* * *

At 6:00 a.m. on Wednesday, 81-year-old Ken Chittum, the long-time volunteer observer at Kerrs Creek, high in the watershed of the James River, called the Weather Service office at Lynchburg to report four inches of rain overnight. The creek was above flood stage and rising fast. Chittum apologized for not calling during the night, but his wife didn't allow him to go out in the storm to check his gauges. Lynchburg teletyped Chittum's and other observers' reports to the Richmond office, who relayed them to the RFC in Harrisburg.[14]

* * *

Wednesday, June 21, 1972
Harrisburg, Pennsylvania

Forecasting responsibilities for the James, Potomac, and other rivers of Virginia were moved from the Washington Weather Service Office to the Middle Atlantic RFC at Harrisburg after the 1969 Camille disaster.[15] As the specialist on the hydrology of those rivers, Michael Mark moved too. All of the hydrologists at the RFC were competent to forecast any of the rivers in their region; the science and protocols were essentially the same. Over time, though, each of them gained a specialty in a particular river basin, getting to know the observers, the strengths and weaknesses in the data, the river's tendencies, and knowledge that comes with familiarity. Mark brought his expertise in Virginia's rivers with him, and retained that specialty at his new post in Harrisburg.

Mark's phone rang at 6:30 a.m. He immediately recognized the measured Tennessee drawl of his Hydrologist-in-Charge, O.D. White. "Sorry for the early call, Mike. Can you get to the office right away?" Mark must have had a sense of *deja vu*, almost the exact words that woke him when Camille hit not three years before.

White continued. "Agnes strengthened to tropical storm over night. Yes, still over land, I've never seen anything like it. With enhanced inflow from the Atlantic. There's already more rain than Washington predicted."

Mark shook his head to clear it. He was well aware that the ground in the headwaters of the James was already saturated. "Jesus. Camille all over again."

"We aren't talking thirty inches," White replied after a pause. "But the rain shield is huge, all of the Blue Ridge and beyond, with at least six inches. Some of the gauges up there are above flood already. It could get bad. We'll need you to run the James and Potomac ahead of schedule, with updates every six hours."

Mark dressed quickly. Images of the destruction wrought by Camille three years before filled his mind. Now, it could be happening again. He sped from his home in Camp Hill to the Federal Building, his windshield wipers set to high.

* * *

Flash floods are local weather phenomena, typically forecast by meteorologists at the local Weather Service office when a storm is imminent or happening *now*. River floods, on the other hand, are slow-moving. It takes time, sometimes days, for a surge in the headwaters to make its way downstream. Hydrologists employ completely different algorithms and models to forecast a flash flood vs. river flood. Accordingly, the Weather Service Offices at Roanoke and Lynchburg issued flash floods warnings for the tributary streams as the storm was happening, while the River Forecast Center in Harrisburg forecast the big James and Potomac Rivers days before the flood arrived, updating constantly.

O.D. White had already started one of the technicians on punching data teletyped from the regional Weather Service offices by the time Mark arrived at the RFC. There were nine gauges on the James River, and several more on the upstream tributaries: the Maury, Cowpasture, Rockfish and Rivanna Rivers. Most of the stations reported in, and Mark had good data to run the model. At 10:00 a.m. he transmitted his forecast to Joe Harden, the Meteorologist-in-Charge at Richmond, to be issued to local response authorities and media. The James, he predicted, would rise to eleven feet at Richmond City Locks by mid-day Friday.[16] Eleven feet, although two feet above flood stage, was nowhere near the record 28-foot flood brought by Camille three years before, and nothing to get excited about.

DO YOU REMEMBER?

Wednesday, June 21, 1972
Scottsville, Virginia

Scottsville, Virginia, a village of five hundred people clustered along the James River sixty miles above Richmond, had seen floods twenty times in the past hundred years. The river sloshed into town so often that it became routine for residents and businesses to move their furniture and merchandise upstairs or to high ground. But Scottsville had never seen anything like 1969's Hurricane Camille, when the river hit thirty feet on the city's gauge, topping a record that stood since 1870. Cars, furniture, and sodden merchandise bobbed about on Main Street as residents escaped their swamped homes by boat.[1]

* * *

Many residents of Scottsville had not yet recovered from Camille when, three years later, the James River began to swell with rains from Agnes. On Wednesday, June 21, 1972, the RFC at Harrisburg predicted that the James River would rise to nineteen feet at Scottsville the next day. By evening the forecast increased to twenty-five feet. *Additional heavy rains since this morning require revision upward for all gauging points in the James basin. ... This is the most serious flood in the James River basin since Storm Camille in 1969.*[2] Still, the forecast was well below previous floods, and not close to the record Camille flood three years before.

The telemark river gauge at Scottsville was two stations above Richmond, and crucial to the RFC's forecast for the capital city. As the river kept rising, surpassing predictions made only hours before, a row of phone poles carrying the communications line to the gauge was undermined by the surging water, and at 9 p.m. collapsed into the flood. The Scottsville gauge went dark. By midnight, seven of the nine upstream gauges on the James River were out of service, swamped in the flood or inaccessible to observers.[3] The hydrologists at the RFC were working half blind. Reports of precipitation mostly continued – ten inches at Scottsville, more than a foot in the mountains. However, without knowing the actual stage of the river at most points upstream, Michael Mark and his colleagues at the RFC were hard-pressed to say when and how high the James River would crest.

Villagers in Scottsville knew the routine, and as they had done three years before and often in the past, lugged their furniture upstairs, piled their goods into trucks, and took shelter with neighbors on high ground. This time, though, the water rose to the ceiling in every home and shop.

At noon on Thursday, June 22, the James River at Scottsdale crested at thirty-four feet,[4] four feet higher than the mark set by Camille three years before. Hydrologist Mark calculated the discharge to be 301,000 cfs – an all-time high, sixty percent greater than the Camille record set just three years before.[5]

* * *

Thursday, June 22, 1972
Richmond, Virginia

The record flood at Scottsville surged down river towards the quarter-million residents of Richmond. With new data in hand the RFC updated its forecast yet again: the James River would rise to twenty-eight feet at Richmond's City Locks – as high as Camille – by 4 p.m. Friday. Joe Harden, a veteran of the Weather Service and Meteorologist-in-Charge at Richmond, issued an urgent warning, aired on the 11 p.m. TV news and all radio stations: *Prepare for severe flooding. Do you remember Camille? Well, we've got it again!*[6]

Richmond City Manager William Leidinger was one of forty on the list of Virginia emergency response authorities notified by the Weather Service when a bad storm was coming. He received the call at his home at 11:30 p.m., his heart racing as he heard that the James would rise to the height reached by Camille, sure to again flood a wide area of the city. Leidinger would get little sleep for next three days.

With the experience of Camille fresh in their minds, Leidinger and the city's response authorities knew what to expect, and what to do. Coincidentally, only two weeks before, Virginia's Civil Defense Command produced an updated emergency plan and held a state-wide dry run of communications, logistics and staffing.[7] They were as ready as they could be.

According to the RFC's forecast, Richmond had almost two days before the worst of the flood hit the city. Without panic, Leidinger set up a command post at Parker Stadium, home of the Richmond Braves minor league baseball team. City officials closed roads and bridges. The Red Cross and Salvation Army opened evacuation centers at schools and armories. City workers and volunteers stacked 50,000 sandbags along Dock and Cary streets to protect water, sewer, and electrical infrastructure. Businesses that flooded in Camille moved their goods and equipment upstairs. 1,800 National Guard troops were activated and arrived at Richmond to help with security. Police urged every resident near the river to get out.[8] With many homes and businesses still not recovered from the wreckage caused by Camille, the people took heed, collected their valuables, and left.[9]

* * *

Shockoe Bottom is one of Richmond's oldest neighborhoods. Founded in 1730 as a tobacco warehouse and shipping port at the upper limit of tidewater, by the late 18th Century it had infamously become one of the largest centers of slave trading in the United States.[10] Most of its warehouses and slave markets were destroyed during the Civil War, then soon replaced by a cluster of Italianate-style homes and shops. Alas, by the mid-20th Century, Shockoe Bottom became isolated by barriers of railroads and elevated highways, and again declined. The warehouses and shops were abandoned, and the neighborhoods marked by poverty and decay.

Over the centuries the James River repeatedly overflowed its banks and flooded into Shockoe Bottom. There was no levee, wall, or other protection from the river. In 1969 the Camille flood brought muddy waters halfway up the first floors of almost every building, and wiped out the businesses in its way. Now, as the Agnes flood rose inexorably higher, many of the shopkeepers, still in debt from rebuilding, could only watch and weep.

* * *

Eleven Virginia Power workers worked feverishly in the after-midnight darkness to protect a high-voltage substation in Richmond's south side. They could barely hear the persistent hum of the big transformers through the pounding rain as they carefully unlocked and opened the disconnect switches to shut the substation down. Public Works men raced to build a sandbag barrier around the facility to keep overflow from the James River at bay, but it was a race they couldn't win. At 2:00 a.m. a section of the sandbag wall collapsed, and water gushed into the substation. Explosions of blue light glinted off the rising water as transformers short-circuited, and all of Richmond south of the river went dark.

The linemen and Public Works men scrambled to the roof of the control building, terrified of a fatal shock if power at the substation electrified the muddy water. They sent a radio message to the fire department, who dispatched a small boat to their rescue. But the boat capsized when the men, all of them, tried to pile in, and were pitched into the flood. The men grappled their way back into the control building, where they stayed, soggy and scared, until another boat came after daylight.[11]

* * *

By daybreak Friday, most of Richmond was sealed off by the flood. Four of the five bridges across the James River, including Interstate 95, were closed when the approaches to them were swamped by the raging James River. Barricades manned by police and National Guard sprung up to keep gawkers and, it was feared, looters, at bay. At 6 a.m. the power failed throughout the city. Downtown Richmond stood silent, dark and deserted. And still, the river continued to rise.

Hampered by failure of upstream river gauges and by sparse reports of actual precipitation, the hydrologists at the RFC almost hourly forecast a higher crest of the James River at Richmond.[12] Yet Joe Harden's urgent bulletin ... *Do you remember Camille?* ... had the effect he intended. Richmonders could not forget the devastation wrought only three years before, and as the forecasts for the flood of the James grew ever higher, the people of Richmond got *out*. Almost incredibly, although there were some very close calls, no lives were lost.

The flood surging down the James from the headwaters gained strength with continued rain and swollen tributaries along the way. Like a slow-motion wave it pulsed through Lynchburg, Scottsville, and Cartersville, exceeding all previous records. Early Friday afternoon, June 23, the James River crested at 36.5 feet on the City Locks gauge, *eight feet* higher than the record set during Camille three years before, literally "off the charts" the Weather Service kept for the river.[13] Discharge topped 319,000 cfs, *more than double* the previous record flow.

* * *

Two hundred square blocks of Richmond – a quarter of the entire city – were swamped as the James River and local tributaries swelled to their record heights. Some of the areas flooded were low-lying industrial parks, rail yards, tank farms, and warehouses, with little residential development. Yet hundreds of downtown businesses, townhouses and apartments also succumbed to the flood. Residents evacuated and survived, but their shops, homes and possessions did not. When the flood receded everything, *everything*, was soaked through with greasy mud, an utter loss. It would be many months, even years, before Richmonders could restore their ruined homes and shops.

Shockoe Bottom buildings flooded to their first floors during Camille flooded to the second floor during Agnes. The few residents who stayed were evacuated by boat. Owners of many of the buildings just gave up and abandoned their property, never to be restored. Agnes dealt a knockout blow to the already teetering Shockoe community.

The Richmond water treatment plant, which every day purified sixty-two million gallons of James River water for the city's 250,000 residents, stood on the banks of the river just upstream from the Powhite Parkway bridge. Backup generators kicked on to keep the plant running when electric power went out, but as the river neared its record crest, even the backup systems succumbed to the rising waters. Richmond, in the midst of its all-time record flood, had no water to drink. Under orders from Virginia governor Linwood Holton, the National Guard swung into action. Twenty 5,000-gallon tank wagons called "water buffalos" were staged throughout the city, and the Army flew in collapsible tanks from a depot in Ohio. Even so, for nearly a week there was barely enough water to drink. There could be no showers or baths, no washing of dishes and clothes, no flushing of toilets. For many Richmonders, the lack of water would be what they remembered the most.

* * *

The devastation wrought by Agnes, followed by another flood in 1985, finally inspired Richmond to protect itself from the James River. In 1994 a system of levees and floodwalls was completed along 3.2 miles of the waterfront. Much of it today has been made part of the city's Floodwall Park, a mecca for hiking, biking, fishing, and outdoor lore, one of modern Richmond's proudest and most prized attractions.

At Shockoe Bottom, left derelict after Agnes, the new flood wall allowed a community renaissance. Abandoned tobacco warehouses became modern apartments, and rundown commercial space transformed into trendy galleries, nightclubs and restaurants. Shockoe's dark history as a slave market is commemorated in a

The James River flooded into Richmond, Virginia, eight feet higher than ever before. (Library of Virginia)

memorial park, and the whole community designated a "National Treasure" by the Trust for Historic Preservation.[14]

Agnes was, and remains, the greatest flood of the James River ever measured, believed to be the greatest flood since colonization of Virginia in 1607. Richmond's new wall was built to hold back a flood reaching twenty-nine feet on the City Locks gauge – high enough to hold back a flood like Camille, but not high enough for another Agnes.[15] It would have been too expensive, and too intrusive, to build that high. The wall has not been truly tested,[16] and Richmonders can only hope they will never again suffer a blow like Agnes.

THE OPERATORS

Tuesday, June 20, 1972
Harrisburg, Pennyslvania

Hurricane Agnes has gasped her last. The Harrisburg Patriot-News picked up the Associated Press story and ran it on the front page.[1] United Press likewise reported *Hurricane Agnes Fizzles out and Dies in Florida*. It sounded like an obituary.

The Patriot-News had been the authoritative voice of Harrisburg since 1852. In those 120 years it never missed a day of publication. Its new three-story brick headquarters on Market Street was six blocks from the river, deliberately sited several feet above the limit of the Susquehanna's record flood of 1936.[2] Reporters could easily walk the few blocks to the capitol and state offices, and to the Federal Building that housed the United States District Court, Selective Service Commission, and River Forecast Center.

The formidable Emma Parker sat vigilantly behind a big desk in the Patriot-News lobby, monitoring everyone who stepped through the front doors. The receptionist/operator, a robust woman about to turn fifty, impeccably dressed and coifed, arrived this Tuesday morning fifteen minutes before her assigned hours. A few minutes later, the owner and executive publisher of the Patriot-News came through the doors. Parker smiled as she handed him a copy of the morning paper, fresh from the press room at the rear of the building. "Good morning, Mr. Baum."

"Good morning, Mrs. Parker, thank you." Most of the staff didn't know her first name, and even the boss called her Mrs. Parker. Baum scanned the front page. The lead story covered state budget negotiations; in Maine, Margaret Chase Smith won the GOP primary for a fifth term as the nation's only woman senator; and, below the fold, the Hurricane Agnes story. The weather blurb predicted *Rain*

Likely Tonight, Clearing Tomorrow. Baum folded the paper and tucked it under his arm. "Will it ever stop raining, Mrs. Parker?"

"It's bound to, sir." Then she added cheerily, "but my dahlias sure do like it!"

The weather column on an inside page gave more particulars. The rain came not from the dying Hurricane Agnes, but ahead of the cold front sweeping down from the Dakotas and stalled over the Great Lakes.

Along with the weather column the Patriot-News published daily reports from the River Forecast Center, including the present and predicted stage of the Susquehanna. At Harrisburg, the stage this Tuesday was five feet, way below "flood stage" of seventeen feet, and forecast to go even lower the next day.

Moments later, one of the cub reporters who covered city politics bustled into the lobby. "Good morning, Mrs. Parker."

"Good morning, Mr. Baer." He wasn't half her age, but Parker treated everyone, from the company president to the press operators, with cordial respect. "You have a telephone message." She tore the little pink memo from its pad and handed it to him, sternly adding "It's long distance," reminding him that outgoing long distance calls could only go thru her.[3]

* * *

Six blocks away, at the River Forecast Center, the daily reports from volunteer observers and other stations were all in hand by 9 a.m. Intern Patsy Quigley took her seat in the windowless computer room at the center of the office to enter the data into the RFC's IBM 1130 computer. Each entry included a 4-digit code identifying the station, *e.g.* Wilkes-Barre, a 4-digit code for the amount of precipitation, and another 4-digit code for the river stage. With reports from two hundred stations it would take her an hour and a half.

The IBM 1130 was 1972's latest in office technology, with 32KB of RAM and a one-megabyte disk drive the size of a pizza. It looked like a standard desk with a keyboard built into the surface. The operator fed data into the machine via a mated punch-card reader, while output printed on continuous tractor-feed paper.[4]

Under O.D. White's direction, hydrologists Mike Gwinner and Michael Mark designed a program in FORTRAN to forecast each of the rivers under the RFC's umbrella. The massive data sets and complex algorithms involved in weather and river forecasting had outgrown the ability of meteorologists and hydrologists to handle manually; now, they could write a program, feed their data in, and allow the machine to run the calculations. It was a wonder the RFC's hydrologists managed their job at all before they had the 1130.

The IBM 1130 computer, with 32KB of RAM and 1MB hard drive, was essential to making weather and river forecasts (Allan Cash Picture Library / Alamy)

It would be impossible, they thought, to go back to paper charts and slide rules.

The objective of river forecasting is straightforward: how high will the river be at a particular place at a particular time? But making such a prediction is highly complex. The greatest variables are the amount and timing of runoff from rain or melting snow. Some of a rainfall will drain to the river, while some will be absorbed by the soil. Factors include how much and how recently it has rained; the composition, depth and slope of soil; atmospheric temperature and humidity; the extent and type of vegetation in the area; and, river flow from upstream and tributaries.

Quigley popped her head into Gwinner's office to say she finished punching the data cards and had stacked them in the reader for entry. The two stepped into the computer room where Gwinner took a seat at the console. He patted the card-reader and looked up at the intern, now standing at his shoulder. Always the teacher, Gwinner asked Quigley if she noticed anything unusual in the data. Anyone could punch data onto cards, but Gwinner wanted to be sure the intern understood what the data meant.

"No, nothing unusual," she replied. "Some of the stations up in New York reported a half inch of rain, but the stages are a little lower than yesterday." The cold front advancing from the Midwest brought showers, but the rain wasn't enough to bring the rivers up.

Gwinner typed "RUN" on the keyboard and ceremoniously pressed the Enter key. The punch cards flipped through the reader as the 1130 received the input, and the printer began chattering on its separate bench. He tore the green and white paper from the device and looked over the numbers. Gwinner had instructed the computer to produce a forecast only for major cities in the Susquehanna basin – Elmira, Binghamton, Wilkes-Barre, Williamsport, and Harrisburg. The hydrologists didn't bother with intermediate stations unless flooding was expected. Later that day hydrologist Michael Mark would run similar forecasts for Virginia's Potomac and James River basins.

Gwinner examined the numbers on the printout, looking for changes between yesterday's forecast and the new one just now calculated. Upriver the predicted stages were slightly higher than predicted yesterday – a result of the rain showers in the headwaters – while here at Harrisburg a little lower. There was still no hint that anything out of the ordinary would come their way.

PILGRIMAGE

Wednesday, June 21, 1972
Lancaster, Pennsylvania

The reborn Tropical Storm Agnes extended its reach into eastern Pennsylvania. Meteorologists called for showers and occasional heavy rain, with possible flash floods on local streams. So far residents didn't worry. A reporter for the Lancaster Intelligencer-Journal even had some fun with it:[1]

> *The Lancaster area faces some potentially heavy rains today from the tattered and bloated remains of Hurricane Agnes. A flash flood watch has been issued where streams are running high from an almost steady diet of recent precipitation.*
>
> *Agnes became a senile old lady Tuesday after battering the Florida panhandle. As she lost her girlish figure, she spread a huge cloud cover over most of the eastern part of the country.*

The article finished with a positive outlook: *All of this should clear out of the area by tonight. Then, there is a good chance of pleasant and sunny conditions Thursday and Friday.*

So very, very wrong.

* * *

Wednesday, June 21, 1972
East Stroudsburg, Pennsylvania

O.D. White left home early to attend a meeting of Pennsylvania's Council on Civil Defense at East Stroudsburg State College. He needed extra time to make a stop along the way. When he reached the village of Analomink, five miles short of Stroudsburg, White checked his map, then drove a narrow dirt road to a cleared area above the banks of Brodhead Creek. Years had gone by since the field was last mowed, and sapling trees encroached along the edges. This was the site of Camp Davis, a summer retreat founded by a Baptist minister from New Jersey.

Back in August, 1955, soon after White took charge at the Middle Atlantic RFC, he was pressed to the limits of his skills by a one-two punch from Hurricanes Connie and Diane. Connie saturated the ground with eight inches of rain, but did not cause major flooding. Then, within a week, Diane roared up the Atlantic seaboard and arced inland to dump ten more inches of rain over Pennsylvania's Pocono Mountains, headwaters of Brodhead Creek, a tributary to the Delaware River.

Forty-six people – mothers, fathers, children – were spending their vacation at Camp Davis. On the night of August 18 they watched from the windows of their bungalows as Brodhead Creek flowed swiftly alongside the camp property, swollen by rain from Diane. The Weather Service in nearby Scranton issued a "flash flood warning" for the Pocono region, but the warning didn't mention Brodhead Creek in particular. A few of the campers, unconcerned, walked through the rain to attend church in town, while the rest of the families settled in at camp for the night.

An hour later, Brodhead Creek, supercharged by continuous heavy rain, rose over its banks to swirl around the foundations of the bungalows. The fragile structures began to tremble as the creek roared around them. Parents piggy-backed their children and ran through the rising brown water to the Big House – the camp's main building – where they thought they would be safe. As they looked back, the bungalows they vacated moments before lifted from their foundations, shattered to splinters, and tumbled down the raging stream. Panicked families scrambled to the second story of the Big House, then to the attic as the creek rose thirty feet in fifteen minutes. Debris-laden flood water crashed through the doors, windows, and floors below as it swirled up the stairs, shaking the whole building. There was no way for the campers to escape when the Big House broke apart and scattered into the flood. Only three of the forty campers who took refuge in the Big House survived.[2]

While the remnant of Hurricane Diane pelted the Poconos, O.D. White, chief of the new RFC at Harrisburg, gathered what information he could from gauges along the Delaware and its tributaries. In 1955 all of the reports came from volunteer observers or local weather stations. There were no automated gauges, nor any computer to run the forecast models. He calculated his predictions by hand, with paper charts of the rating curves and a slide rule to do the math. Using algorithms he personally devised, White predicted record floods – a "double crest", one after another – on the Delaware River from Stroudsburg all the way down to Trenton. He was spot on, and riverside communities were warned in time. There was no question that White's forecasts saved lives. His success in forecasting the Connie/Diane floods on the Delaware River brought White fame in the weather community, and gave a boost to development of river forecast centers around the country.

Back in the present, 1972, White walked silently around the broken foundation of the Big House, musing how far river forecasting had come in the last eighteen years, yet how far they had to go. In 1955 there were no stream gauges or weather stations on Brodhead Creek, and no one could have predicted the rain would be heaviest on this particular stream. With the tragedy at Rapid City ten days ago fresh in his mind, and White's own history with the flood that destroyed Camp Davis, the crucial importance of his profession was palpable. River forecasting was, truly, a matter of life-and-death.

White returned to his car and drove the next few miles to the conference at East Stroudsburg. Several dozen officials from Pennsylvania's Council on Civil Defense attended, chaired by its *ex officio* director Lieutenant Governor Ernest Kline, along with representatives of other government agencies and civic organizations. General Townend was there from Wilkes-Barre, and exchanged greetings with White when he arrived during lunch hour. These meetings were held every year to bring the people who stood at the front lines of disaster response together.

After lunch, White gave a presentation about the latest protocols for getting flood forecasts out to emergency responders. He concluded his remarks with an *ad lib* warning. "Even now, what's left of Hurricane Agnes is causing floods in the South. There's a good chance we'll get some of that here." White couldn't know the turns Agnes would take, but he did know that the ground was already saturated, that rain from the stalled cold front was adding to that load, and that the path of a tropical storm such as Agnes was not perfectly predictable.

Nobody understood flooding in the region better than White. "So we could be needed at our stations, and it may be best if this meeting ended today." A murmur spread through the room. The conference was supposed to continue the next day. Then, perhaps with the surprise disasters at Camp Davis and Rapid City in mind, White paused before concluding in a measured tone. "It's coming, and it's going to be big."[3]

* * *

Wednesday, June 21, 1972
Corning, New York

Dr. Robert Brill neatly folded his clothes and tucked them into a suitcase for his trip to Afghanistan. A chemist by training, Brill was the science director at the Corning Museum of Glass. The trustees agreed to sponsor his trip to investigate a historic glass factory in the remote city of Herat, where artisans crafted exotically colored glass using techniques handed down for centuries. Brill was excited about seeing it with his own eyes; he was, after all, a leading expert – no, *the* leading expert – on the composition and manufacture of ancient glass.

Corning Glass Works, with its headquarters and main factory here in Corning, founded the Museum of Glass in 1950, and built it into the foremost repository of glass art and artifacts in the world. Tourists visited from all over the United States, and around the globe, to view the treasures. In addition to glass exhibits, the museum housed a priceless library of books and documents about glass art and manufacture, including rare manuscripts from as early as the 14th century. The library was the only place in the world that scholars could study these one-of-a-kind archives.

The museum and library were housed at the Corning Glass Center, a two-story extension of the Corning Glass Works headquarters a hundred yards from the beautiful Chemung River. The river flowed serenely this day, only a few feet deep. Brill never saw it flood, although he heard the levees and walls along the riverbanks were built after a big one back in 1946.

The television was on in Brill's bedroom as he packed. News reports showed flooding in Virginia, caused, the newsman said, by the remnants of Hurricane Agnes. He didn't pay much attention. Richmond, after all, was 350 miles away, and hurricanes were of no concern in upstate New York. Still, he noticed that the intermittent rain that plagued the region was pattering more heavily on his windows. The rain came with the stalled front sagging from the Midwest; it was not, yet, due to the remote remnant of Agnes. Turning again to his suitcase, Brill carefully wrapped a Steuben crystal vase and tucked it between his folded clothes, a gift for his host at Harat.

It would take two days for Brill to reach Afghanistan: a short hop from Syracuse to JFK in New York, a red-eye to Amsterdam, a connection through friendly Tehran, and the final leg to Kabul. From there, after an overnight in a hotel, a driver would take him several hundred miles to the ancient glassworks.

New York Route 414 wound through the famous Finger Lakes wine country between Corning and the airport at Syracuse. Normally, a couple of small waterfalls cascade beside the road as it descends to the village of Watkins Glen. But this day the falls enveloped the entire face of the cliff and splashed over the pavement. Brill had driven this road dozens of times, but had never seen the falls so wide and powerful.

SUPERSTORM

Wednesday, June 21, 1972
Washington, DC

The National Hurricane Center in Miami resumed its lead role in forecasting the renewed Tropical Storm Agnes when the storm center swirled back over the ocean near Virginia Beach. Director Bob Simpson requested a Navy reconnaissance plane out of Jacksonville to take the measure of the reborn storm. To his surprise, the flight reported a central pressure of 977mb off the Delmarva Peninsula, lower than recorded even when Agnes was a hurricane in the Gulf of Mexico. Steady winds were back up to sixty-nine mph – not quite a hurricane, but a very strong tropical storm.[1]

At the same time, a separate center of low pressure formed along the cold front sagging in from the Midwest. Days before, storms ahead of this front panicked residents of Rapid City, South Dakota, still digging out from the deadly flash flood that struck the previous week. Now, in a complex atmospheric *pas de deux*, the twin lows spun along in parallel, with the resurgent tropical storm Agnes flinging Atlantic moisture into the new vortex to its west.[2] Agnes became a monster, what would later be known as a "superstorm".[3]

The rapid re-development of Agnes, and its interaction with the new center of low pressure to the west, caught meteorologists by surprise.[4] Observations from weather stations, volunteer observers, balloon soundings, radar, satellite imagery, and other data weren't coming in fast enough to forecast the storm's unexpected strength and path. To make matters worse, many of the surprises came at night, when observations were scant and Weather Service offices not fully manned.

Howard Hoover, Meteorologist-in-Charge at the Weather Service Office at Washington, struggled to staff his office around the clock to keep up with the reborn tempest. The day before, Agnes

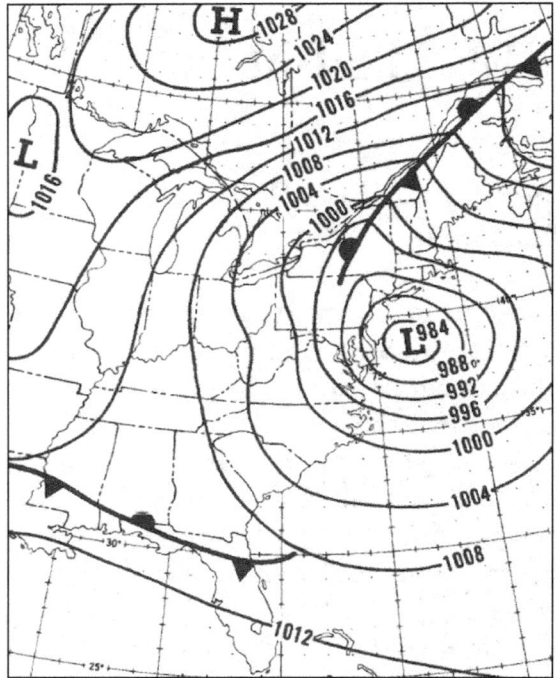

Surface weather map June 22, 1972, depicting the reborn Tropical Storm Agnes off the New Jersey coast and its interaction with a cold front advancing from the Midwest (NWS)

was expected to continue its eastward path and dissipate over the Atlantic, and he forecast intermittent showers in the DC metro area. But now, as Agnes curved to the north and merged with the sagging cold front, its enormous cover of rain engulfed the entire Middle-Atlantic region. Flash flooding was almost certain, and it was all hands on deck at the Weather Service to stay on top of the threat. The meteorologists and technicians rotated shifts though day and night to gather data, run the forecast models, and alert first responders, the media, and the public.

The mushrooming of the disaster, and the Weather Service's near-panic to keep up, is revealed in the increasing urgency of bulletins issued by the Washington office.[5] The first flash-flood warning came at 3 p.m. Wednesday afternoon, June 21, almost routine:

Flash flood warnings are in effect for tonight for Fairfax, Prince William, and Arlington counties in Virginia, and for the following counties in Maryland: Montgomery, Prince Georges, Howard, Baltimore, Harford, Cecil, Carroll, Frederick, Washington, Allegany.

The situation got worse, fast, with the next bulletin at 7 p.m.:

Much of the Virginia area has now received 5 to 6 inches of rain this afternoon, and indications are that the Maryland area could experience the same amounts of rain this evening. All interests in this area should be advised that this is a major flood situation.

The 10 p.m. update:

Major flooding with many roads closed, substantial damage and evacuation of residents in many areas from continuing torrential rains. At Dulles Airport two more inches of rain has fallen in the last two hours, bringing the storm total to nearly 8 inches.

At midnight Washington began issuing updated bulletins every hour, now including frantic superlatives, as the deluge continued.

1:00 a.m., Thursday, June 22:

Many reports of extreme difficulty to persons and property have been received. High water poses a serious threat to life when only two feet deep and moving swiftly. Ask yourself – do you know how deep the water is?

2:00 a.m.:

Reports of loss of life ... Extreme difficulty to persons and property ... Fallen trees and power lines ... Washed out bridges ... Closed roads ... Persons venturing forth do so at their own risk and almost any reason is not sufficient for travel.

3:00 a.m.:

Rainfall amounts of 7 to 11 inches have been reported north and west of the city, and 5 to 8 inches east. Dulles Airport recorded a 24-hour total of 10.83 inches. ... These totals are over or near all time records.

4:00 a.m.:

Major flooding continues in the area. Wholesale evacuation of thousands has and continues to occur. Damage of major disaster proportions has occurred.

It was not, yet, the great Potomac River bringing the "disaster of major proportions" to Washington and its suburbs. Rather, it was the local creeks and little tributaries that swelled with the unprecedented rain, rising quickly over their banks to record levels.

Streams in Northern Virginia are often named "runs." At Alexandria, in view of the Washington Monument and Capitol Building, little Cameron Run and its tributaries drain a 44-square-mile watershed inside the Capital Beltway The watershed is almost entirely urbanized, and there isn't much forested land to slow runoff from heavy rains. A dam halfway up the run forms Lake Barcroft, surrounded by a private community focused on lakeside amenities.

Within two hours after issuing the first flash-flood warning, the Weather Service Office at Washington phoned the Alexandria police to urge immediate evacuation of people living along Cameron Run.[6] At 8 p.m., fireman Larry Jenkins was dispatched with four others to help with evacuations and rescues. An hour later, for the first time since it was built in 1915, water crested over Barcroft Dam, eventually three feet over, sending a sudden wave rushing downstream. Jenkins and his comrades had just escorted an elderly couple from their creek-side home when the surge surrounded them. "All of a sudden, it just threw us up against the house and we couldn't move," Jenkins recalled. Other rescuers saw Jenkins' predicament and launched a small rowboat, tethered to a rope, for the men to climb into. But the boat capsized in the torrent and tumbled away. Desperate, the rescuers threw a line to the firemen, who this time managed to grab hold. "Hand-over-hand, we came out of the water, and we were pretty horizontal. I came so close to dying that night".[7] Cameron Run crested at 18.14 feet, four feet higher than ever before, a height never since approached.

Four Mile Run, which courses past the Pentagon in Arlington, raged far higher than its record crest and collapsed the Walter Reed

Drive Bridge into torrent below. Thousands of residents of Arlington and Alexandria evacuated to shelter in nearby school buildings or were stranded in their creek-side homes.

At Manassas, Virginia, little Bull Run winds through a National Historic Park commemorating two horrific Civil War battles. Communities of suburban Washington encroached on the old battlefields, and the tension between development and historic preservation remains palpable to this day. In nearby Chantilly, sixteen inches of rain fell during June 21-22, flooding Bull Run along the edge of the battlefield and into the surrounding communities. The Loch Lomond area, where water rose to the rooftops of two hundred homes, was especially hard hit.

The swollen waters of Bull Run raged downstream to join the Occoquan River just above the 55-foot-high dam holding back the Occoquan Reservoir, the source of fresh water for much of Northern Virginia. Officials feared the dam might collapse, and ordered everyone in the village of Occoquan – 975 people – to get out *now*. Late Wednesday night the river spilled over the top of the dam, ultimately by eleven feet, to inundate the town and destroy everything in its path.[8]

Flood water spilling over Occoquan Dam swamped the pump station below and knocked out water service for a half million residents of Northern Virginia.[9] Engineers drew some fresh water from nearby Washington and Arlington, and stationed twenty big tank trucks around the county. Residents lined up to get five gallons of fresh water per person at local fire stations, but had to bring their own bucket. To make matters worse, overflow from sewage treatment plants made its way into the water supply, raising fears of contamination. Fairfax County offered free tetanus and typhoid shots to any resident who wanted them.

The VA-123 bridge over the Occoquan River, a steel truss standing since 1875 and the main route through the village of Occoquan, shook with the violence of the rampaging river then collapsed into the torrent. The skeletal girders of the crumbled old bridge pitched downstream until they crashed into the piers of the "Purple Heart" Bridge carrying Interstate-95. Virginia DOT ordered the big bridge closed, fearing it destabilized by the impact.[10] Farther downstream the flood chewed at the piers of the high bridge carrying US-1 until they too gave way. The entire south-bound span ripped apart and fell sixty feet into the river. All of the main roads leading south from Washington – I-95, US-1, and VA-123 – were cut off.

The southbound side of the US-1 bridge over the Occoquan River collapsed into the flood (Library of Virginia)

LONE WOLF

Wednesday, June 22, 1972
Washington, DC

Flooding on the Maryland side of the Potomac was as terrifying and destructive as suffered in Virginia. Fifteen inches of rain fell at the Carroll County town of Westminster, 9.5 inches of it on June 22 alone – the all-time state record for rain in a single day. More than ten inches fell over most of Carroll, Frederick, Montgomery and Baltimore Counties.

At Laurel, the Patuxent River crested at twenty-five feet: fifteen feet above flood stage, six feet higher than ever before. The discharge of 26,000 cfs more than doubled the record established only a year earlier.[1] The river swelled into downtown Laurel, swamped the businesses along Main Street and US-1, and swept away the historic Ninth Street and Race Track Bridges. At least a thousand residents evacuated to high ground.[2]

The Monocacy River at Frederick, Seneca Creek at Dawsonville, and the Northwest and Northeast Branches of the Anacostia River reached all-time and still-standing record discharges, multiples of the previous peaks, causing catastrophic damage to homes and businesses nestled along their banks. Roads were scoured and littered with debris, bridges washed away, and signals deadened without power. Traffic in Washington's suburbs came to a standstill.

* * *

In Washington, D.C., Rock Creek Park is a serpentine green oasis through the heart of the city. The rough-and-tumble stream running the length of the park is paralleled by Beach Drive, a twisting two-lane road with turnouts to picnic groves, playgrounds, historic Pierce Mill, and the National Zoo. Commuters use the bucolic route to avoid the stop-and-go traffic on DC's major thoroughfares, even though the drive might be longer and slower.

Rock Creek was known to occasionally flood, but with little development in the park there was never much damage. The worst flood occurred in 1956, when the creek rose to thirteen feet on the Geological Survey gauge – six feet above flood stage – with a discharge of 7,220 cfs.

On the night of June 21 a handful of late commuters, plus revelers returning to their suburban homes after dinner or a show, came to a sudden halt on Beach Drive where the rising water of Rock Creek lapped over both ends of the historic Boulder Bridge. The drivers made U-turns to backtrack, only to find the creek swamping the approach to the Blagden Road Bridge they safely crossed minutes before. Exiting their cars, the travelers were trapped. Strangers, still in business or dinner dress, huddled to plot their escape. Some wanted to stay and wait for rescue, or for the water to recede. But with the roaring creek lapping at their feet and rising, their only course was to scramble through the dark woods up and out of the valley to the Northwest DC communities above. Several dozen cars – including four police vehicles – were swamped, some of them carried downstream, but all of the drivers and passengers made it to safety. At its peak the next morning Rock Creek ran three feet higher than ever before or since, its flow almost double the previous record.[3]

* * *

A little further downstream Rock Creek swept alongside the National Zoo. Most of the animal exhibits stood up the hill and away from the flood. Only the zoo's maintenance shops and a temporary animal holding pen were near enough to the creek to be in danger. At the height of the flood these buildings would be swallowed by eight feet of muddy water.

Zoo director Theodore Reed arrived at his headquarters shortly after 9:00 p.m., alerted to flash flooding throughout the region and worried about Rock Creek. Reed had fostered a revolution in zoo management here, built expansive habitats in place of confining cages, and initiated a breeding program for endangered species. Just two months earlier Reed brought Ling-Ling and Hsing-Hsing to his zoo, the first giant pandas to arrive in the United States, a signature gesture in thawing relations with China. Reed's delight in and love of his animal charges was revealed by the tigers, bears and leopards he raised at his own suburban home.[4]

Reed was encouraged by the reports from his diligent staff. The pandas were secure in the indoors portion of their habitat, and the big cats moved to safe shelter. Two of the four elephants decided to stay out in the rain, standing in the lee of their big concrete shed but in no danger from the rising creek.

The holding pen below the maintenance shops housed a mother timber wolf and five pups born a month ago, out of the public eye until the pups could tolerate the zoo crowds. The pen was too close to Rock Creek, so assistant zookeeper Amanda Howard took the mother and four of the pups to an empty behind-the-scenes cage uphill. "But the other pup didn't follow and I couldn't find her. We've got to go back down there!"

Reed jumped up. One of his animals was in danger. "Let's go!" He and assistant keeper Howard sped down the zoo road to the wolf pen and braked to a stop. A swirling eddy from the raging Rock Creek flooded the pen and most of the parking area behind the maintenance shops. As Reed and Howard stepped from the truck they heard the whines of the stranded pup piercing through the relentless drum of the rain. Reed shone his flashlight into the pen and found the young wolf standing atop the lean-to shelter that had been her litter's den. The lean-to was five feet high, but only a few inches of its peaked roof remained above the roiled water. The pup saw his rescuers and yelped loudly.

"Hold this." Reed handed the flashlight to Howard. Without saying more he stripped off his shirt and held it in his hand, waded into the dark water, unlocked the gate, and strode toward the whimpering young wolf.[5] The flood was up to Reed's armpits, and the swirling current, loaded with debris, nearly knocked him from his feet.

Howard kept the light trained on the shelter, shouting encouraging words to Reed and the terrified wolf pup. The animal knew her voice well – this was the human who tended the litter every day – and calmed as Reed reached the flooded shelter, more afraid of the water than she was of him. Reed wrapped his wet shirt around the pup, which weighed only ten pounds, and held her close to his chest as he struggled back to the vehicle. He handed the pup to Howard, who clutched the shivering animal in her arms, cooing and scratching behind its ears. When they released the pup into the temporary cage to join her litter-mates, the mother wolf nuzzled and groomed her pup profusely – and, it seemed to Ted Reed and Amanda Howard, gratefully. Reed turned to his assistant. "What do you think, Amanda? Should we name her Agnes?"

When the rain stopped the next morning, keepers were astonished to discover a near-miracle. At the height of the storm two rare red pandas were born in the hollow of a Sycamore tree, the first red panda births ever at the National Zoo, or anywhere in America since 1954.[6]

Damage from the storm kept the zoo closed until the following Tuesday. Until then, Reed proudly posted a sign near the zoo entrance at Connecticut Avenue: *All our animals are safe! See you again soon!*

* * *

For three days the great Potomac River at Washington ran chocolate brown with silt washed from fields and farms upriver. The Army Corps of Engineers mounded sandbags around the Jefferson and Lincoln Memorials to guard against inundation. The flood scoured away sixty-six miles of the old towpath in C&O Canal National Park, and the network of steel footbridges to the overlooks at Great Falls fell like pick-up sticks.[7] Even the White House was not immune from Agnes, as the incessant rains made a pond around the mansion and invaded the press room in the basement.[8]

The flood of the Potomac peaked early Saturday morning at 16.7 feet on the gauge near Washington's Wisconsin Avenue, five feet above "major flood stage" and near the record St. Patrick's day flood of 1936.[9] If the peak came at high tide it would have been the highest flood of the Potomac in history.[10]

EAGLE SCOUT

Thursday, June 22, 1972
Ellicott City, Maryland

Little Ellicott City, in a steep valley of the Patapsco River fifteen miles west of Baltimore, played a prominent role in the industrial history of America. Founded in 1777 when brothers John, Andrew and Joseph Ellicott built a grain mill along the river, the city became one of the largest manufacturing centers in early America. Its products were so vital that a new way was found to get them to market; in 1830 Ellicotts' Mills dispatched the first commercial shipment by rail in the United States. The original Baltimore & Ohio depot, built in 1829, the oldest in America, still stands on high ground near the river. Enriched by their mills, the Ellicott brothers built magnificent stone mansions in the valley.

The Patapsco River at Ellicott City is a double-edged sword. The river provided power for the mills, but sometimes disastrous floods roared down the narrow valley.[1] The deadliest, and perhaps the strangest, occurred July 24, 1868. There was no rain in Ellicott City itself, but a series of thunderstorms upriver unleashed a deluge into the Patapsco Valley. Dark clouds loomed to the west, and flashes of lightning zig-zagged across the angry sky. Residents of Ellicott City braced for rain that never came. But the flood did come, as the Patapsco rose ten feet to lap at the doorsteps of the homes in the valley. Then, suddenly, there was a terrible roar, as what witnesses described as a wall of water rushed down the river into the city.[2] Twenty-eight homes and two of the mills were destroyed, and forty-three residents lost their lives. The freak flood of 1868 remained the highest on the Patapsco for 104 years. And then came Agnes.

Just before midnight, Wednesday, June 21, 1972, police and firemen fanned out to alert the residents of Ellicott City and other Patapsco Valley towns of the flood warning issued by the Weather

Service. The river was rising unbelievably fast, and unbelievably high. Most people escaped up the hillsides to find refuge in evacuation centers at local schools and churches. By noon Thursday, virtually every building in the valley at Ellicott City was swept away, and many more ravaged in nearby Elkridge and Sykesville. The 200-year old John Ellicott mansion, built of massive stone, ripped in half as it was pounded by debris. General Edwin Warfield, leader of Maryland's National Guard, toured the scene by helicopter. "I've never seen anything like this in my life. Every structure within 100 feet of the Patapsco along a twenty-mile stretch was either carried away by the rampage, or destroyed where it stood."[3]

Seven people lost their lives to the Patapsco.[4] A couple of boys plunged over a dam and drowned in the whirlpool below when they tried to raft down the raging river. Not far upstream from Ellicott City two young women leaving the popular Hollofield Inn didn't make it across the parking lot before flood waters carried them away.[5]

For every death there were many heroic rescues. Firemen in Ellicott City risked their own lives as they ferried families away from second-story apartments in small boats.[6] Upstream, near Marriottsville, the flood stranded two engineers of a 137-car coal train when rising waters engulfed the riverside tracks. Just after a helicopter plucked the men from the cab of their locomotive, the embankment gave way and the train tumbled into the river.[7]

* * *

Thursday, June 22, 1972
Columbia, Maryland

During the summer of 1972, Kathy McCardle, a student at the University of Pennsylvania, worked as a waitress in Columbia, Maryland. Twenty-year-old Tom Girven was enjoying dinner with his family at Kathy's restaurant when she caught his eye, and he hers. Before long they struck up a whispered conversation, and he shyly asked if he could call her.

They set a date for Wednesday evening, June 21. Despite the incessant rain Girven arrived early at McCardle's apartment. They were soon off to as nice a dinner as Tom could afford, followed by the 10 p.m. showing of The Godfather at the new Columbia City Theater.

It was almost 1 a.m. when the movie let out. McDardle and Gervin were overwhelmed by the intense, dark saga played out by Marlon Brando, Al Pacino and the rest, and didn't talk much as they walked to Gervin's car. Not that they could have heard each other over the rain pounding on the pavement and cars around them.

McCardle's apartment was on the east side of U.S. Rt. 29, the main artery through Columbia, while the business district was on the west. To take her home, Gervin drove down South Entrance Road alongside the Merriweather Post Center for the Performing Arts. He braked to a stop where the road crossed Little Patuxent River. Although his headlights barely penetrated the rain, Gervin could see water lapping over the road ahead of him.

Little Patuxent River is normally a babbling brook that connects a series of man-made lakes in the planned city of Columbia, passing under South Entrance Road at its intersection with U.S. 29. But this night it wasn't babbling. It was roaring.

"Look at that! We can't drive through it. What are we doing to do?"McCardle didn't seem nervous, but enjoyed the unexpected adventure. By now a few other cars pulled up behind them, none of the drivers foolish enough to ford the raging stream.

Gervin sensed McCardle's excitement. "We'll have to go back and cross 29 at the North Entrance. No problem." This would have been true. "But this is amazing. You want to watch it for a while?" It seemed like a good time to put his arm around McCardle's shoulder, and she scooted across the front seat to sit against him.

At this hour the rainfall in central Maryland accumulated at more than an inch an hour. Every drop of it that fell over the saturated watershed of the Little Patuxent ran right into the creek, now rising a foot every ten minutes. Gauged that night at Savage, a few miles downstream from Columbia, the Little Patuxent reached a peak discharge of 32,500 cfs, more than *five times* the previous record.[8]

Gervin and McCardle stayed to watch a few minutes too long.

Lost in the excitement of the storm, and of their closeness to each other, they didn't realize that the flood over the road was lapping at the wheels of their car. Gervin told the story the next day.[9] "Suddenly the water rose and took the car into the current. It was filling up with water. We got out and stood on the doorsills. The water was going so fast it was impossible to walk out."

A fireman called to the scene roped up and tried to reach the stranded pair, but lost his footing in the current and barely managed to save himself. The chief called the Marriotsville Fire Department, eight miles away, for their rescue boat. There was nothing more they could do before the boat arrived. It wouldn't get there in time.

The water kept rising. "It was spilling over the roof and onto our heads." After half an hour clinging to the car, Gervin and McCardle could no longer hang on. "She let go and I grabbed her. It was pitch black. The water was rough. Kathy got panicky and let go, and we were separated."

Gervin was carried half a mile downstream until swept into a tree. "I was ready to black out. Another two seconds and it would have been all over." He climbed up and held onto the tree, swaying wildly in the floodwaters, the rest of the night. By daybreak the flash flood began to recede, and Gervin, an Eagle Scout and trained lifeguard, decided his best chance was swim for safety. "I managed to make it, maybe 50 or 100 yards. I was shaking and it took a while to get myself together." With flood waters receding and debris piled everywhere, Gervin didn't know where he was. "I walked down the road and found a cop, and rode around with him for a couple of hours. I am more concerned about Kathy."

McCardle' body was found that Thursday afternoon, entangled in trees not far from Columbia's South Entrance Road.

Howard County, bounded by the Patapsco River to the north and the Patuxent River to the south, became a veritable island. Of the thirty-nine bridges that spanned those rivers, only three – the high bridges of Interstate 95 and the U.S. 40 over the Patuxent – remained passable.[10] Many were washed away or damaged beyond repair. Victims and other residents couldn't get out, while rescue and recovery teams couldn't get in.

I CAN'T SWIM

Thursday, June 22, 1972
Harrisburg, Pennsylvania

The River Forecast Center was staffed 7 a.m. to 5 p.m. most days, but when their rivers ran high one of the hydrologists worked late, even overnight. Now, with so much unexpected rain from Agnes, Mike Gwinner stayed until midnight Wednesday to receive reports from observers, compile the data, and get a head start on running the forecast models the next morning.

Gwinner arrived back at the Federal Building Thursday morning after only four hours sleep. Several dozen poncho-clad protestors marched on the sidewalk at the front doors, the painted letters on their placards bleeding in the rain. City police kept just enough space open to allow Gwinner's entry. He popped his head into O.D. White's office when he got up to the RFC. "What's going on out there?"

"Can't miss them, can you? It's lottery day for Selective Service." All across America young men anxiously awaited their draft numbers; would they have to go to Vietnam, or not? Even in the rain the protesters were there to picket at the Federal Building.

* * *

That same morning, John Baum, executive publisher of the Patriot News, shook a spray of water from his umbrella as he snapped it closed inside the front door of his newspaper building. As usual, receptionist Emma Parker greeted him and handed him the morning paper, just off the press in the rear annex of the building. She wore a pastel yellow dress with a magenta dahlia pinned near her shoulder. Baum gestured toward the flower, "From your garden, Mrs. Parker?"

Parker beamed. "Yes sir, I picked it this morning. Yesterday was the first day of summer, you know!"

"Well, they must enjoy this rain. But it seems to me we've had enough." Baum glanced at the weather blurb on the front page. The Patriot-News took its forecast from the Weather Service Office at Harrisburg's airport: *Rain today, heavy at times. A flash flood warning remains in effect this morning.* The lead article reported Sen. George McGovern's victory in the New York primary, assuring his nomination as the Democratic candidate for President.

"Yes Mr. Baum. My bus went through some pretty deep water this morning." Storm sewers in parts of Harrisburg couldn't cope with so much rain. "I hope I can get home after work."

"I am sure it will be alright. It's supposed to clear up by tonight." Baum smiled at Parker, tucked the newspaper under his arm, and stepped to his office at the rear of the lobby.

Parker's phone rang as soon as Baum stepped away. It was her duty as the lead receptionist to route calls to the appropriate news desk, department, reporter, or other staff. "Good morning, Patriot News. May I help you?"

The caller stammered nervously. "I want to report ... there is a flood. I live on Orrs Bridge Road, in Camp Hill. Cars are stuck in it. It ... it looks like the people can't get out. And it's almost up to my door." The Patriot-News, like many local papers, encouraged citizens to report newsworthy events.

"Have you called the Fire Department, ma'am?"

"No, I didn't think of that. But I want to report it, so called the paper."

Parker's phone never rang as much as it did that morning. Citizens called to report roads flooded, cars abandoned, bridges washed out, homes isolated by the rising water. There were desperate pleas for help from people stranded in their homes, and from witnesses to rescues. All of the Patriot-News reporters were out covering the unexpected flood story, while the editors, print-setters and press men scrambled to revise, then re-revise, the Evening News to be published that afternoon. John Baum himself ran to and fro from his executive office to the editorial floor upstairs, and to the press floor at the rear, trying to manage the fast breaking story. All Parker could do was take the names and phone numbers of the citizen callers. "Yes ma'am, I will see to it that your call is returned as soon as possible. You stay safe, do you hear?"

Word spread throughout the city, and through the offices of the Patriot-News, that the Susquehanna was rising quickly, already far above the level predicted by the River Forecast Center the night before. John Baum assured his staff – who were becoming anxious about their own homes and families – that the offices of the Patriot-News were safe from the big river. Their building, he said, stood well above the Susquehanna's record flood.

What Baum did not anticipate was an attack from the rear. Paxton Creek, only fourteen miles long, sprouts in the wooded uplands east of Harrisburg, meanders down through pastures and suburbs, then is confined to concrete channels and culverts meant to keep flood waters out of nearby rail yards and industries. The culvert ran only a block to the east of the Patriot-News.

At 10 a.m., a reporter came in to write up his notes, and passed Parker on his way to the editorial floor. "It's getting really bad out there. Paxton Creek is up over the road. Don't count on getting out of here tonight." Parker only vaguely knew the creek was there; its channel was hidden by overhanging trees. She walked to the front door to look outside, and gasped to see a foot of water flowing fast down Market Street into a small lake under the railroad overpass two blocks to the west.

At 11 a.m. John Baum directed the administrative staff to put their records into boxes and carry them upstairs. Financial, circulation, and personnel records were saved first. Baum especially worried about the press room itself, located on the first floor to the rear of the building. The huge, very expensive, presses couldn't be moved, and would be ruined if water got into the building.

Paxton Creek kept rising as the deluge continued. It overflowed its concrete culverts, coursed down Market Street, and lapped up the steps to the front door of the Patriot News. Just after 2:00 p.m. water came into the press room from the loading dock at the rear. Disheartened, but with no choice, John Baum ordered that power to the presses be disconnected. The great machines whined to a halt. For the first time in 120 years, the Patriot-News would fail to publish.[1]

Paxton Creek rose higher, into the building. By 5:00 p.m. water covered the desks on the first floor, destroying typewriters, office machines, and furniture. $250,000 worth of newsprint in the pressroom annex was ruined. The presses themselves were half-submerged, their precision parts and electrical connections beyond repair.

Emma Parker's dahlia hung limply from her yellow dress. She looked as wilted as her flower, glistening from the toil of packing and toting boxes of records to the second floor. She peered anxiously from an upstairs window to see muddy water rising up to the windows of cars abandoned on Market Street. The over-flowing Paxton Creek formed a relatively calm pond, six feet deep, around the Patriot-News building, while only a block away a white-water torrent raged toward the Susquehanna.

No one knew how much higher the flood could go, or how much battering the building could stand. John Baum deemed it too dangerous to stay, and called the fire department to evacuate the dozen employees still on hand. Overwhelmed by rescues throughout the city, the department could spare only one small rowboat. Four or five people at a time crammed into the little boat to be ferried to high ground near the state Capitol. An hour went by before the boat returned from delivering the first group to safety, and another hour for the second group. The flood continued to rise. Only John Baum, Emma Parker, and two other employees remained. It was nearly dark when the exhausted fireman rowed into sight and nosed the boat to the front door.

"Alright Mrs. Parker, it's our turn. Are you ready?" Baum extended his hand to his receptionist.

Parker hesitated, her voice quavering. "Mr. Baum, I can't swim."

Baum glanced at the brown water in the murky twilight. It would be dark soon. "You won't have to, and I'll help you. It will be fine. We have to leave. Now!"

They sloshed thigh-deep through the lobby, past floating wooden desks, a coffee table, and water-logged papers. Parker, in stocking feet, held her high heels in her left hand and tightly gripped Baum's hand with her right. They waded out the front door, and Baum guided her to the seat at the stern of the boat.

Moments after the fireman shoved away a strong current through the alley alongside the building spun the little boat sideways. When the boat collided with a telephone pole Parker pitched over the stern and into the flood.

"Emma!" Baum leapt to his feet, almost capsizing the boat. By now she was eight feet away. The fireman reached for her with the oar. Panicked, Parker got her hand on it, but despite the encouraging shouts of the passengers couldn't hold on. Parker went down, then re-emerged sputtering brown water. "Swim, Emma, swim to the boat!" She went down again. The fireman rowed furiously after her – away from safety – until they came to the edge of the raging white-water at the channel of Paxton Creek. They could go no further. Emma Parker was gone.[2]

* * *

Six blocks away from the drama unfolding at the Patriot News, O. D. White clamped his unlit pipe firmly between his lips. For the first time in the history of the RFC every one of the rivers on his watch was in major flood, and still rising. All of his hydrologists, technicians and administrative personnel were on hand, their hours extended to assure 24-hour coverage. Mike Gwinner again volunteered to stay to all night.

Reports came in from observers and telemark gauges every three hours, data cards were punched, and the IBM-1130 ran constantly to update the river forecasts – every time predicting higher floods than the previous run. According to protocols, the forecasts were transmitted to Pennsylvania's Council on Civil Defense and the regional Weather Service Offices, then from there to local media and response authorities.

Just before noon Patsy Quigley made contact with her counterpart Weather Service intern at Binghamton, New York, via the RFC's big radio console. Binghamton collected reports from observers and automated gauges in the New York portion of the RFC's territory – the Chemung and North Branch of the Susquehanna Rivers – then transmitted them as a batch to the RFC. This spared the RFC staff the time required to contact the twenty stations in New York directly. There was no banter on the transmission this day, but strictly hurried business. As the Binghamton intern read the data – codes for station ID, stage, and precipitation – Quigley penned it onto the printed form from which she would punch it onto cards for the next model run. Just after the Binghamton intern finished transmitting the data from the North Branch and began reading the codes for stations on the Chemung, the radio emitted a shrill squeal. Then, nothing but static. Quigley tapped the transmit button "Binghamton, do you read me?" There was no response.

A 300-foot radio tower stood near the Appalachian Trail at the summit of Peters Mountain a few miles north of Harrisburg. The tower served as a dedicated repeater for emergency transmissions and other government business, including Weather Service communications. Just as Patsy Quigley began to receive the Chemung River data, a 50-mph gust of wind whipped rain under the shields protecting the equipment atop the tower and penetrated the electronics. The short circuit tripped a breaker, and the transmitter went dead. Radio communications between the RFC and the Weather Service Office at Binghamton, as well as to all other stations, were cut off.

O.D. White telephoned the Meteorologist-in-Charge at Binghamton and arranged for further transmission of the New York data via teletype. At least those lines were working. Loss of the radio was not an ultimate disaster, as the RFC had protocols for alternative communications. But teletype would be slower and take more of the RFC's time, one more straw on the camel's back.

* * *

In August, 1933, Codorus Creek at York, Pennsylvania, on the west side of the Susquehanna, flooded at 32,000 cfs after a coastal storm inundated the area. At least one person in the city drowned. To protect York from another disaster – and with funds from the 1936

Flood Protection Act – the Corps of Engineers built Indian Rock Dam just west of the city.

Agnes proved too powerful for the dam. Fifteen inches of rain filled the reservoir until, for the first and only time, it came over the top in a controlled spill, bringing Codorus Creek to its record height.[3] Five residents of York drowned, including a young man swept off the Richland Avenue Bridge while he tried to take pictures of the flood.[4]

* * *

In Lancaster County, to the east of the Susquehanna, rains from Agnes swelled the Conestoga River to twice its record discharge. Six of the county's iconic covered bridges, the oldest dating back to 1843,[5] ripped apart and washed away.

Near the village of Paradise, Amishman Samuel Kaufmann drove his buggy along Frogtown Road, his strong hands tight around the reins. It was almost impossible to see the road through the rain-streaked window. His wife Sarah sat anxiously beside him with their baby Benjamin cradled in her arms. Pequa Creek, running alongside the road, sloshed over the pavement. "With God's grace, Sarah, we'll make it. Just a few more minutes."

A car slowed behind the buggy, and the driver flashed his lights to indicate he wanted to pass. Kaufman's horse, expertly trained and normally compliant, whinnied and balked as Kaufmann guided him to the right. Suddenly the front wheel of the buggy dipped into a hidden wash-out, and the whole affair rolled sideways into the creek. The horse broke free when the rig twisted, but the buggy tumbled with the current until it lodged against a tree. Brown water quickly filled the cab. Sarah managed to keep her head up, and was pulled to safety by firemen called to the scene. But her husband Samuel and baby Benjamin did not survive, among the eight residents of Lancaster County lost to Agnes.[6]

* * *

At nearby Chadds Ford, Pennsylvania, curator John Sheppard and his staff raced to move priceless paintings by Andrew and J.C. Wyeth from the galleries of the Brandywine River Museum of Art. At its height[7] the river rose halfway up the first floor, turning the museum into a sort of inside-out aquarium "We saw fish and frogs swimming right by our museum, where the water was eighteen inches up the windows," Sheppard told a reporter.[8] Sheppard saved all of the art, but the museum remained closed for weeks while volunteers mopped out the mud and restored the galleries.

RATS!

Thursday, June 22, 1972
Harrisburg, Pennsylvania

"Rats!" Muriel Shapp screamed when she saw the dark rodent scramble across the grand hall of Pennsylvania's Executive Mansion.[1] The thirty-one-room Georgian-style edifice, at the north end of Harrisburg just yards from the Susquehanna River, was built only four years earlier. Governor and Mrs. Shapp's living quarters were upstairs, while the first floor featured the governor's private office, a huge formal dining room, and galleries of Pennsylvania art and antiques.[2] Mrs. Shapp much preferred her own home near Philadelphia;[3] their very public life in the mansion, she said, felt like living in a fish bowl. Now, she faced a flood rising outside the front door, and, worse, rats. She detested the place.

The state trooper assigned to the mansion ran from his security office when he heard Muriel's scream. "Mrs. Shapp?"

"A giant rat!" She gestured toward a hallway where her dog Cleopatra stood barking, her ears back. "Right there!"

"It probably came up from the basement." A rat was the least of his concerns. "There's a foot of water down there already. Come here, look."

The guard opened the big front door, with Pennsylvania's first lady at his side, and peered out over the courtyard. Rain pelted down harder than Muriel Shapp had ever seen, thick droplets blown by the gusty wind into undulating sheets. The air itself seemed liquid. Overflow from the Susquehanna River sloshed against the steps and washed over the brick pavement and manicured gardens. Only the points of the wrought iron fence along Front Street stood above the swirling brown water. If the river rose two feet higher, as now predicted, the brown water would come over the doorsill and into the mansion.

Police cadets, interns from the Department of State, and a curator from the State Museum arrived to help the maintenance staff move the mansion's expensive furniture, antiques and art up to the living quarters.[4] Maintenance men hefted the heavier pieces, while the security guard took personal care of the governor's official papers. The curator fussed as the interns ineptly dismounted oversize paintings from the walls and rolled up the plush Persian carpet in the dining room. To and fro, up and down the stairs, they toiled all afternoon. By the time they finished, flood water came up through the basement and puddled throughout the first floor. Only the heavy grand piano and draperies remained.

Muriel's husband, Governor Milton Shapp, returned to the mansion from his office in the Capitol at 10 p.m., entering through a service entrance at the rear. He had been leading his government's battle against the disaster unfolding across the state all day, and would get no rest this night. Shortly after midnight flood water rose over the sill of the big front doors and surged through the first-floor public areas. Electric power blinked out, and water taps went dry. Shapp told Muriel to pack an overnight bag for a stay at her brother's Harrisburg apartment.

Two state troopers arrived at the mansion in a police boat at 3:30 Friday morning[5] to evacuate Governor and Mrs. Shapp. Misty rain drizzled around them, but not the pelting deluge of the day before. Flood water in the rear parking area, four feet deep, was relatively calm, and the troopers easily guided their small boat to the service door. Muriel waded through the kitchen in brown water up to her knees to the waiting boat, bereft to leave Cleopatra behind with the security detail. Wet, unkempt, and exhausted, she stepped aboard and slumped in the seat next to her husband, her slicker clutched tightly around her body.

* * *

Harrisburg was at the center of the maelstrom. Twelve and a half inches of rain fell at the airport Wednesday night into Thursday, and 15.25 inches overall. By Thursday morning all of the creeks and

Pennsylvania's Governor Milton Shapp and his wife Muriel were evacuated when the newly-constructed executive mansion flooded to the ceiling (Elmira Star-Gazette/USA Today Network)

little rivers that fed the Susquehanna around Harrisburg ran higher than their all-time record floods, and kept rising. On the west side Yellow Breeches, Conodoguinet, Fishing, and Sherman Creeks; on the east side, Wiconisco, Big and Little Swatara, Clarks, Asylum, Spring, and Paxton Creeks swelled to double, triple, five times their previous record flows.

At the U.S. Department of Agriculture's Mahantango Research Station, in the rolling hills thirty miles north of Harrisburg, technicians measured 18.2 inches of rain between Wednesday, June 20 and Friday, June 23.[6] Although there were anecdotal reports of gauges overflowing and rainfall up to twenty-two inches in other places, Mahantango was the maximum total rainfall officially recorded anywhere during Agnes. The flow of Mahantango Creek reached 7000 cfs, *seven times* the prior record.

The combined and almost instantaneous runoff from up to a foot-and-a-half of rain around Harrisburg led to a local anomaly. Fed by extreme flash flooding on dozens of local tributaries, the Susquehanna rose to twenty-six feet on the gauge at Harrisburg Thursday night, nearly as high as the 1936 St. Patrick's Day flood. When the locally extreme rain abated after midnight, the stage of the Susquehanna began to trend downward. City authorities thought the worst was over. They were wrong.

* * *

Governor Shapp enjoyed very little sleep at his brother-n-law's apartment. He arose early Friday, got an update from aides, then set out on a tour of the flooded areas of Harrisburg. Escorted by the state police commissioner, they took the same small boat used to evacuate the mansion earlier that morning. The Susquehanna would not reach its maximum crest until Saturday, but already fifteen percent of the downtown and residential areas of the city were under water. On some streets the flood ran sixteen feet deep. The helmsman swerved to avoid street signs and phone poles, and the party ducked beneath overhead wires and traffic lights.[7] When they approached the Executive Mansion, Shapp saw flood waters halfway up its grand first floor windows.

Governor Shapp made a rescue of his own that Friday morning. He not only wanted to inspect damage at the mansion and collect some personal items, but Muriel begged him to retrieve her beloved dog. As the state police escort nosed the little boat up to the front doors, two of the troopers on guard at the mansion waded through the grand lobby, one carrying two suitcases and the other with the whimpering Cleopatra cradled in his arms. She had been well cared for by the security detail, but was keen to be reunited with her family and wagged her tail when Governor Shapp took her into the boat.[8]

* * *

Thirteen row houses exploded in flames two blocks from the Governor's Mansion (Courtesy of Bryson Leidich Photography)

An air-shattering boom shook the boat and its passengers as they motored away from the Executive Mansion, so loud they feared it came from the mansion itself. Sparked by static electricity, gas leaking at a row house on Penn Street, two blocks from the mansion, ignited to send a ball of flame and smoke high above the city's rooftops. Fragments of wood and brick flew in all directions. Within minutes an entire row of thirteen homes was engulfed in flames.

Firefighters raced to the scene, but were stymied when they drove their trucks into flood water higher than the tires. Some waded and swam the final three blocks, tugging their heavy hoses behind them, while others perched on nearby rooftops to train their hoses on the blaze.[9] Helicopters thrummed overhead to rescue residents who hadn't evacuated. Fortunately, everyone escaped unharmed. All thirteen homes burned to the waterline. When the flood receded, only their gutted brick shells and piles of ashy debris remained.

* * *

The following week, when the river receded enough to allow entry to the Executive Mansion, maintenance staff found the high water mark only two feet below the ceiling of the first-floor galleries. A huge dead carp lay sprawled in the formal dining room, and refugee rats scurried through the hallways. Two inches of greasy mud coated everything. The plaster walls sagged and hardwood floors buckled. Mold infested the extravagant draperies, the water-logged wood of the grand piano swelled to ruin, and nineteenth-century hand-blocked wallpaper in the grand entry blistered and crumbled. Two years would go by before renovations allowed Milton and Muriel Shapp to return. To this day, a small brass plaque, affixed high on the wall in the grand foyer, reads simply "1972 Hurricane Agnes."[10]

* * *

The DeHart Dam, built in 1940 of earth and stone in the hills north of Harrisburg, held back Clarks Valley Reservoir to provide drinking water to the city and its suburbs. Less than a mile downstream from the dam stood the rustic buildings of Camp Shikellimy, run by the Greater Harrisburg YMCA. Campers, many from poor sections of the city, planned to spend a week at Shikellimy, hiking in the forest, making crafts, holding bonfires. But this Thursday night, June 22, there would be no bonfire, as a hundred campers and twenty staff huddled in the dining hall to avoid the driving rain.

The Harrisburg Water Authority opened the spill gate on DeHart Dam earlier that day to relieve pressure on the dam and allow the excess flow to escape downstream. The Shikellimy campers heard the rush of Clarks Creek as it overflowed its banks and tore at Clarks Valley Road. Within thirty minutes the creek washed out the bridge into camp and cut gullies forty feet wide through the only road out of the valley. The campers were trapped. Worse, the chief engineer for the Water Authority feared that DeHart dam might be overtopped or breached to send a wall of water down Clarks Creek Valley and into Camp Shikellimy.

What started as an adventure became desperation. The YMCA director contacted the state police, who called the local commander of the Pennsylvania National Guard. The Guard dispatched two CH-47 "Chinook" heavy-lift helicopters to Shikellimy to rescue the stranded campers.[11] As the pilots set the big choppers down on the ballfield, the campers and staff ran from the dining hall through the pelting rain to be boosted aboard by the guardsmen. From there it was a twenty-minute flight to safety at the Harrisburg Armory. City kids who expected a week of crafts and campfires instead had peril and rescue, the experience of a lifetime. DeHart Dam held, but weeks passed before Clarks Valley road re-opened and Camp Shikellimy could again be reached.

* * *

Thirty-five miles north of Harrisburg, Little Gravel Run swelled with the relentless, record rain falling over the nearby hills. The culvert for the normally trivial creek beneath the Penn Central Railroad's tracks along the river wasn't nearly big enough to carry the flood, and by midnight the raging stream chewed a huge gully through the grade. The rails remained, suspended in mid-air over the wash-out.

Jim Kovars was jolted awake by screeching metal and a rumbling crash at 2:30 Thursday morning. It could only be a train wreck; the Penn Central line ran just a quarter mile from his farm. Kovars rousted his brother Richard, and the two men walked along the tracks through the pounding rain to see if they could help. They arrived at the scene of the wreck in minutes.[12]

Engineers Irv Grover and John Buttles had been driving a train of four locomotives and seventy freight cars from Wilkes-Barre to Altoona. The blinding rain limited their view ahead, and they never saw the washout. When the lead locomotive broke through the suspended rails it plunged head-first into gully, pulling the trailing three locos and four freight cars with it. Grover and Buttles weren't injured, but found themselves trapped in the cab with muddy water rising around them.

The Kovars brothers called to the engineers through the roar of the rain and flood, urging them to climb along the wreckage to

A Penn Central freight train crashed through a washout near Sunbury
(Penn Central Post)

the edge of the washout. Grover and Buttles were pulled to safety, though soaked and rattled, and spent the rest of the night at the Kovars' farm. It would be several days before work trains could get to the site to clear the wreck and repair the line, as bridges were washed out both north and south.[13] The Gravel Run wreck was only the beginning of the existential disaster that Agnes would bring to the region's railroads.

BUCKET BRIGADE

Thursday, June 22, 1972
Douglassville, Pennsylvania

The Schuylkill River begins high in coal country above Minersville, Pennsylvania, not far from the center of maximum rain measured at the Mahantango Research Station. The extreme deluge that fed west-flowing tributaries to the Susquehanna River around Harrisburg also supercharged the headwaters of the Schuylkill, which flows 135 miles eastward to join the tidewater Delaware River at Philadelphia.

Flash floods ripped into the little cities of New Philadelphia, Port Carbon, Pottsville, and Schuylkill Haven on the upper reaches of the Schuylkill Wednesday night. When the surge reached the old industrial city of Reading Thursday morning the river swelled to an all-time record stage, five feet higher than the mark set back in 1850. Thousands of drums of hazardous chemicals, some ruptured, floated from shipping docks and storage lots to sweep forty-five miles down the river to Philadelphia.[1] The historic Cross-Keys bridge lifted from its piers and wrenched away, and the venerable Gibraltar Bridge crumbled. Panicked residents fled to shelter at Daniel Boone High School when the rising flood encroached on row houses along Carpenter and 4th Streets. They crowded there, sweaty and scared, subsisting on Red Cross rations, until the water receded.

* * *

At Douglassville, twelve miles below Reading, Lester Schurr heard the flash flood warning for the Schuylkill when he turned on his radio Thursday morning. His family's home on Main Street stood three hundred yards away from and twenty feet higher than the river, well above flood stage. He didn't worry about a flood at his home, but Schurr was worried about Berks Associates, the business he managed in the industrial section of town.

Berks Associates, which collected and recycled crankcase oil drained from cars and trucks, had been in operation since Schurr's father founded the company thirty years before. They recovered ninety-five percent of the waste oil and sold it as a lubricant or ingredient of asphalt. The remaining sludgy residue was stored in two huge earthen lagoons behind the Berks recovery plant.[2] Just two years before, heavy rain filled the lagoons and breached the surrounding berms to release the oily sludge into the river. The cost of the cleanup, lawsuits, and fines imposed by the government bankrupted the company. Schurr was convinced that the trauma of the spill contributed to his father's death only two months later.[3] The company came out of bankruptcy, and under the watchful eyes of state and federal regulators built reinforced berms around the waste lagoons.

Now, Lester Schurr faced another disaster. There seemed little danger his improved berms would breach, but the continuing heavy rain might once again fill the lagoons and carry some of the sludge over the top.

Schurr's dozen employees began operations at the plant Thursday morning as usual, but as the rain pounded down he turned their attention to protecting the property. The crew moved their trucks and lifts to higher ground, and with a makeshift plow mounded soil around four hundred drums of reclaimed oil in a bid to protect the valuable product. At 11 a.m., as the river lapped at the doors to the plant, Schurr sent his anxious workers home. The power failed just after noon,[4] so he locked the plant and drove a mile up Main Street to join his own family.

The Schuylkill River was forecast to rise to a record flood by Thursday evening, almost certain to invade Schurr's home. From the time he arrived home until late in the evening, Lester Schurr, with his wife and two children, frantically carried everything they could from the first floor to safety on the second. At its crest the water filled their basement and came two feet up the walls of the first floor. It was bad, but the Schurrs saved most of their furniture, appliances and carpets. Many residents of the valley would suffer much worse.

All emergency services in the Schuylkill Valley were fully engaged in evacuation, rescue and protection of property. The Coast Guard, which had primary responsibility to recover from oil spills, sent most of its force to collect drums of chemicals washed into the Schuylkill estuary at Philadelphia.[5] An oil spill at Douglassville was not on anyone's mind.

The swirling water of the Schuylkill River rose over the berms of Berks Associates' lagoons late Thursday afternoon. No one was there to see it happen. The new berms held, but the river climbed ten feet over the top. All eight million gallons of oily sludge thought safe in the lagoons flushed down the river.[6]

The Schuylkill began to retreat Friday morning. Schurr's plant foreman noticed a sheen on the roiling river as he made his way to the plant, the first sign there was a spill. He immediately called Lester Schurr at his home, where Schurr and his family were busy shoveling mud out of their kitchen. "Les, there's oil on the river, and I think it's from our lagoons."

Schurr set his muddied shovel down. "Are the dikes okay?" The foreman assured him they stood strong. The river was still over its banks, but receded enough to see the berms intact. There was no breach as there had been two years before. Still, some of the sludge could have floated out with the flood. To make matters worse, the drums of product they tried to protect were gone. "I'll get down there in half an hour."

Even the slightest release had to be reported, so Schurr called the Environmental Protection Agency at Philadelphia. EPA inspector Malcolm Castor flew up the river by helicopter to inspect the damage. As his flight approached Pottstown, Castor saw patches of black oil on the river and spreading into the country on both sides. The oil spill was a catastrophe on top of a major disaster.[7]

For seventeen miles downriver from Douglassville the spill coated riverbank trees twenty feet high,[8] in a distinct line separating leafy green above and black sludge below. The goo pooled in forests and farmland, and infiltrated every structure in the flood's path.

"We've been through floods before," said Paula Townsend of Pottstown, her eyes burning from the oil's acrid fumes.[9] "But before, we could wash everything down with a hose. Now we have all this oil, and you can't get it off."[10]

The Coast Guard had experience with oil spills in bays and on beaches, but not in forests, farmlands and neighborhoods. "We've never had a spill in flood conditions before," said EPA spokesman Alan Jennings.[11]

EPA Administrator William Ruckelshaus visited Pottstown to view the mess a week later. "We have the finest teams in the world working here," he said. "But it's not a simple question of an oil spill on a beach. It will take a lot of time and imagination to clean this up."[12]

Cleaning up the oil-soaked countryside indeed stretched the imaginations of the Coast Guard, EPA, and their contractors. They first employed tried-and-true vacuum trucks, excavators, floating booms, high-pressure hoses, straw matting, and absorbents. None of these methods, good perhaps on a beach, were adequate. In the end, "we went back to the bucket brigade technology of the early Egyptians,"[13] said EPA regional director Ed Furia. Youths worked twelve-hour shifts, for a good wage of $3.50 per hour, to shovel sludge into pails, then hand the pails down the line to waiting dump trucks. Where sludge penetrated deep into woods not accessible by truck, Amish farmers brought their horses and mules to sledge out barrels of oily soil.[14] The cleanup continued until the end of September, but it would take years for the riverside trees to grow back.

Lester Schurr and Berks Associates were not blamed for the spill. The company's reinforced dikes didn't fail. No one anticipated the Schuylkill River could rise high enough to sweep over and scour out the lagoons. EPA Administrator Ruckelshaus called the flood an unpredictable disaster, an "Act of God".[15] The Douglassville spill, eight million gallons of oily sludge, was and remains today the worst inland spill in U.S. history.[16]

MAN OVERBOARD

Thursday, June 22, 1972
Pottstown, Pennsylvania

Below Douglassville the Schuylkill River roared past Royersford, Phoenixville, Norristown, and Conshohocken all the way to Philadelphia, wreaking havoc to riverside industries and neighborhoods. Roads and bridges were washed out, water and sewage plants overcome, utilities shut down. In the words of the Philadelphia Inquirer, the river became "a writhing, hundred-mile-long monster."[1]

The gauge on the Schuylkill River at Pottstown, four miles below Douglassville, was overcome Thursday afternoon when the river hit twenty-two feet, breaking the record set in 1904.[2] Only one gauge on the entire river, at Reading, remained in operation. The River Forecast Center, hampered by the scantiness of the data and unable to account for rain yet to come, predicted that the Schuylkill would climb to a frightening twenty-six feet Thursday night. The prediction was wrong: the river rose to thirty feet – four feet above the RFC's dire forecast, and nine feet higher than ever before.[3] The discharge swelled to 95,000 cfs, double the prior record. Nevertheless, warnings were timely issued to police and civil defense, and most residents of the Schuylkill Valley managed to get out of harm's way.

The industries at Pottstown clustered along the river, and there was no way to protect them. Bethlehem Steel, NL Industries, Columbia Boiler Works, and Mayer-Polluck Steel flooded several feet deep, their equipment ruined. The landmark Mrs. Smith's Pies factory, maker of dessert pastries served throughout the Northeast, didn't stand a chance. Bakers became truck drivers and moved the company's fleet of delivery vehicles to high ground and safety, As at many riverside towns, rainwater gushed in from storm sewers

The Schuylkill River at Pottstown rose nine feet higher than ever before (Philadelphia Bulletin/Temple University Libraries)

and overwhelmed Pottstown's sewage treatment plant. A million gallons of untreated sewage spewed into the Schuylkill.[4]

Neighborhoods presumed safe from the river were not. At 2 a.m. Thursday morning, having heard the forecasts, Pottstown police officer Ed Hoffman roused his family from their sleep at their River Road home and told them to pack a bag. They had to leave, right now. Then he went next door to alert his neighbors, a pair of elderly sisters. The sisters had seen floods before, and told Hoffman they would stay. At daybreak, with his family safe at his brother's home up the hill, Hoffman went back to again urge the sisters to leave. But again they declined, so Officer Hoffman stayed with them. As the flood rose into the house the three went to the second floor. From there, with muddy waters rising up the stairs, Hoffman helped the sisters through a window and onto the roof of the front porch. A rescue boat came by, but the women were too frightened to climb aboard. Then, a helicopter. With the house shaking violently and the swirling water lapping at their feet, Hoffman finally convinced the terrified sisters that they *had* to go. One by one he helped them

into the basket lowered from the chopper, and they were hoisted up
to safety. Hoffman himself was the last lifted out. "He stayed with
them for almost twelve hours," marveled Pottstown mayor Broward
Yerger. "He did a tremendous thing."[5]

* * *

Thursday, June 22, 1972
Philadelphia, Pennsylvania

At 11 p.m. that Thursday night, Officers Luke Lanahan, Leo
van Winkle, and Joe Johnson of the Philadelphia Police Department[6]
were dispatched to rescue a woman stranded in rising water near
the mammoth Container Corporation plant in the Manyunk section
of the city. The historic plant stood on Venice Island, separated from
the main stem of the Schuylkill River by the Manyunk Canal. The
veteran officers, members of the Motor Harbor Patrol, were all
strong swimmers, trained and experienced in water rescue. They
were working their second shift of the day, after evacuating workers
from several of the riverside factories.

The officers planned to rescue the stranded woman first, then
secure a 6000-gallon tank truck loaded with oil at a nearby factory.
Lanahan released their motor launch from its trailer and slid it into
an eddy by a concrete dock at the foot of Nixon Street. The roiling
river glistened in the dim illumination of streetlights along the
Schuylkill Expressway on the other side. Dark logs, oil drums, and
debris appeared and disappeared like ghosts as the churning water
raced past. The men gave the straps on their life vests an extra tug
as Lanahan took his seat at the rear and started the motor.

The motor died only seconds after it started, but the current
had a grip on the little boat and pulled it toward the center of the
river. Lanahan frantically pulled the starter cord as the boat turned
broadside and pitched up a huge wave. The craft capsized as the
wave broke back on itself and tossed the three policemen into the
river.

Lanahan fell close to the riverbank, and fought his way back to the dock where he was pulled out by a bystander. Frantic, he scanned the dark water for his comrades. When he couldn't see them he sloshed to the police truck and radioed headquarters. "This is Harbor Patrol Lanahan. Two officers in the river! Johnson and van Winkle. We capsized just off Nixon Street!"

Within minutes, dozens, then over a hundred, police officers and staff swarmed the riverbanks and bridges to hunt for the lost officers. They commandeered private cars along East and West River Drives and directed the drivers to turn their headlights toward the Schuylkill. Two helicopters chattered overhead to train their powerful searchlights onto the dark, frothy water. Radios buzzed with possible sightings, and warnings.

Officer Johnson kept his head up, buoyed by his life jacket and clinging to a piece of flotsam. At the Columbia Bridge, two miles downstream from where the boat capsized, a searcher spotted him and tossed an inner tube tied to a rope. Johnson held onto the tube, but the rope slipped off. The rushing current pulled him further down river. "If he goes over the dam, he's done for!" At Fairmont Dam, at the head of the tidal Schuylkill estuary, the river drops eight feet into a turbulent, deadly trap for anything that falls over it.

Near the Girard Street Bridge, three-quarters of a mile – just minutes – above Fairmont Dam, Harbor Patrol officers in another boat spotted Johnson bobbing in the current. They grabbed him, but didn't dare hoist him aboard for fear their boat would capsize. Instead they dragged Johnson behind as they motored to the west bank. Other police and civilian bystanders formed a human chain to haul Johnson up the muddy slope through vines and briars to a waiting ambulance.

"This has been the worst day of my life," Johnson told his rescuers as he collapsed from exhaustion. "I am going to have that life jacket framed!"

The search continued through the night for Officer van Winkle. A life jacket found near Peters Island apparently wasn't his. There was otherwise no sign of him. Hearts sank as his fellow officers realized their comrade had probably gone over Fairmont Dam, almost certain death. Four days later, commercial boaters found the body of Leo van Winkle ten miles downstream from where he fell overboard, well below Fairmont Dam and not far from the Philadelphia Navy Yard. Officer Leo van Winkle, dead at 45, left behind his wife and three young children.[7]

* * *

Observers recorded "only" 3.5 inches of rain at Philadelphia, a hundred miles east of the heaviest precipitation. Still, the massive surge from upriver pushed the Schuylkill higher than anyone could remember. On Friday afternoon discharge of the Schuylkill at Philadelphia peaked at 103,000 cfs, in second place behind the record established in 1870. Although most of downtown Philly was sited on high ground, the surging Schuylkill swamped riverside streets and industries, and caused millions of dollars worth of damages.

Thousands of drums of toxic and explosive chemicals swept off loading ramps, storage yards and industrial plants at Reading, Douglassville, and Pottstown floated forty-five miles down the Schuylkill all the way to Philadelphia.[8] There they crashed over Fairmont Dam to pile up with broken trees and other detritus in a backwater at the foot of the Philadelphia Art Museum.[9] The reeking debris clustered so thickly it seemed a person could walk on it. Over the next several weeks the Coast Guard and Philadelphia Fire Department, carefully, very carefully, captured the drums of volatile chemicals one by one, then towed them to a riverside wharf to await disposal.

EPICENTER

Wednesday, June 21, 1972
Hornell, New York

Violent winds, pummeling rain and roaring floods took dozens of lives and caused millions of dollars of damage as Agnes raged up from Florida through the Carolinas, Virginia, Maryland and southern Pennsylvania. The storm inflicted a major disaster over five states. Then, when the rain diminished to misty drizzles, residents thought the worst was over. The remnant of Agnes, forecasters said, was headed out to sea to dissipate over the Atlantic. Sunny days were ahead.

Agnes had other ideas. The storm and floods to the south were but prelude to the unimaginable catastrophe she was about

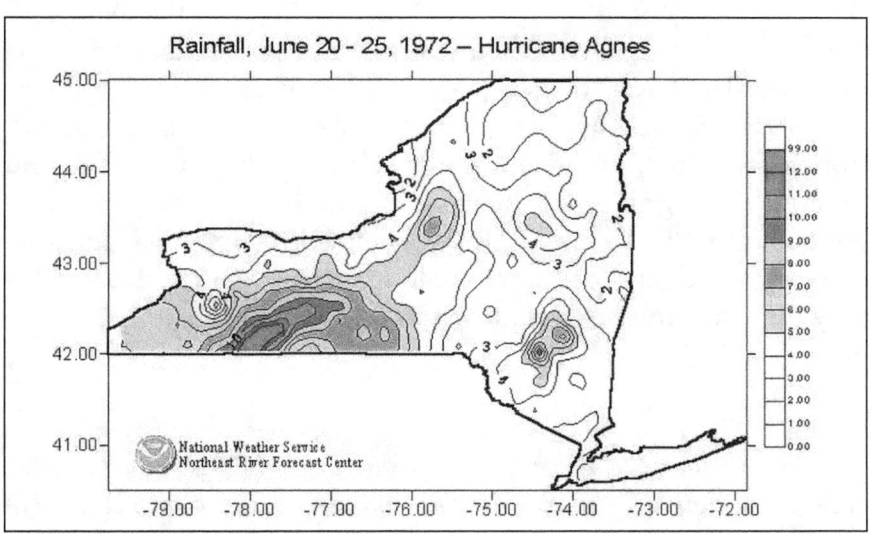

The epicenter of extreme rainfall – up to 16 inches – in New York's Southern Tier (extending south into Pennsylvania), headwaters of the Allegheny, Genesee and Chemung Rivers (NWS)

to unleash in the Susquehanna River basin. Fed by an epicenter of extreme rainfall over the "Twin Tiers" region of western New York and north-central Pennsylvania, the worst of the Agnes floods caught forecasters and riverside communities by surprise in a 400-mile rampage from the Finger Lakes of upstate New York to Maryland's Chesapeake Bay.

* * *

The rolling landscape of the Twin Tiers is sparsely populated, home to a handful of isolated villages. Four great rivers begin in this remote region: the Allegheny west to Pittsburgh; the Genesee north to Lake Ontario; the West Branch of the Susquehanna to the south; and, the Chemung River – a major tributary of the great Susquehanna – to the east. In some places, the headwaters of these rivers are only yards apart, yet are separated by the "Eastern Triple Divide", flowing into the Mississippi, Great Lakes, and Atlantic watersheds. As such, predicting the flow of these rivers falls under the jurisdiction of different River Forecast Centers: the Ohio RFC handles the Allegheny, the Northeast RFC the Genesee,[1] and the Middle Atlantic RFC the Susquehanna and its tributaries.

Between Tuesday, June 20 and Thursday, June 22, a one-two punch of violent, drenching rain pummeled the Twin Tiers, and all four of the great rivers reached their record flows. The floods beginning on the Chemung River wreaked the most havoc of all, bringing death and destruction along its own course then all the way down the mighty Susquehanna.

* * *

The weather forecast for the Twin Tiers on Tuesday, June 20, called for *cloudy with showers and thunderstorms, a chance of rain tonight, cloudy periods tomorrow.*[2] The predicted showers came from the cold front advancing from the Midwest; forecasters did not yet understand the role Agnes had to play in the unfolding atmospheric drama.

"Cloudy with showers" proved to be a drastic understatement. The re-strengthened Agnes inhaled moisture from the Atlantic when it moved off the Virginia capes, only to exhale it hundreds of miles inland. When this warm, sodden air collided with the cold front over the Twin Tiers it fell as an incredible six to eight inches of rain.[3] Flash floods raged into Coudersport, Pennsylvania, and Almond, Hornell, Wellsville, and Olean, New York overnight. Emergency operations got underway only when the few village police on night duty noticed water washing over creek-side streets. By 4 a.m. Wednesday, police, fire, civil defense and volunteers frantically awakened the owners of village shops and urged them to move merchandise and equipment up from their showroom floors.[4] Basements flooded, bridges washed out, roads closed. Authorities told riverside residents to shelter at local churches.

On Wednesday morning the Weather Service office at Boston predicted that the stalled mid-western front would soon wring itself out, while the remains of Agnes would dissipate over the ocean. Residents of the Twin-Tier villages relaxed as it appeared the worst was over. An 8:30 a.m. bulletin from Rochester predicted the rain would end by noon.[5] And the rain did stop, although scowling clouds scudded by. The headwaters of all four major rivers – near or above their previous record highs – stopped rising by early afternoon,[6] and villagers prepared to clean up the mess. The show is over, folks.

The show had barely begun.

No one imagined the next trick Agnes had up her sleeve. Steered by upper level winds and pulled into the orbit of an extra-tropical Low along the stalled front, Agnes made an unexpected turn westward and moved over Pennsylvania. "We didn't anticipate it would move so far inland", marveled Silvio Simplico, director of the Weather Service's Eastern Region. In fact, the Weather Service didn't see the storm's turn until hours after it happened. Forecasters "stretched to the limits of human endurance" predicted the remnant center of Agnes would take a northward path up the Hudson Valley, and were baffled when hours later it appeared over central Pennsylvania. "The storm was so erratic that it was unpredictable."[7]

The very core of Agnes stalled over the Twin Tiers, carrying its prodigious load of Atlantic moisture. Every stream in the region was already flooding, and the hillside soils were utterly saturated. The pump was primed for unthinkable catastrophe. "Agnes stayed over us for two more days and rained herself out, literally, right over the Twin-Tier region. And *that* is what really killed us."[8]

Rain began anew on Wednesday afternoon, a thick, steamy, relentless rain produced by the still-beating heart of Agnes[9]. *Another* six to eight inches fell in the next twenty-hours,[10] added rain that could not soak into the soil, could not evaporate, but could only swell the flow of already-raging rivers.[11] As NBC-TV anchorman John Chancellor intoned on his broadcast that evening, "more rain is forecast for tonight and tomorrow, *and the worst could get worse.*"[12]

Meteorologists and hydrologists at the Weather Service Offices and River Forecast Centers couldn't keep up with the burgeoning disaster. Every river gauge in the epicenter was overtopped or washed away. Every volunteer observer fled to high ground or couldn't be reached because their phone lines were down. There were few reliable real-time reports of the renewed rainfall in the headwaters of four major rivers. Making matters worse, the epicenter

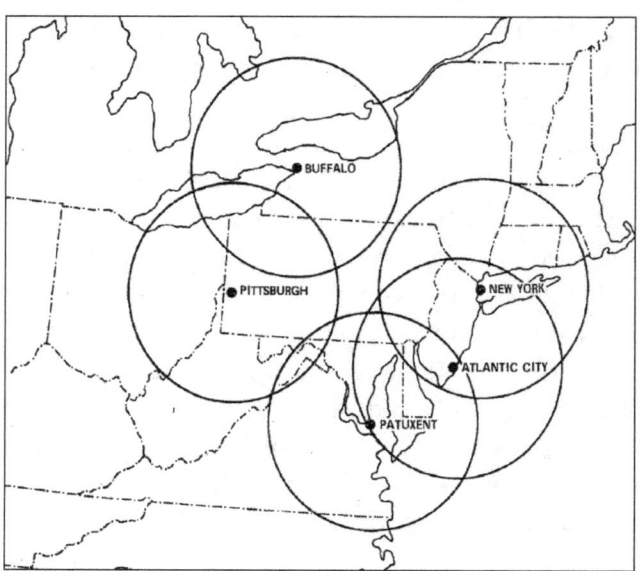

Radar coverage was sparse over the epicenter; operators could see that it was raining, but could not see how much (U.S. Army Corps of Engineers, Post-Audit Agnes Floods)

was on the very fringe of weather radar at Buffalo and Pittsburgh – the operators could see it was raining, but couldn't see how much.[13] Forecasters were nearly blind, and the villagers of the Twin Tiers had little warning of the catastrophe that was upon them.

GRAND CANYON OF THE EAST

Wednesday, June 21, 1972
Rochester, New York

The Genesee River rises in the Allegheny Plateau of New York's southern tier, courses over dramatic falls in the "Grand Canyon of the East" at Letchworth State Park, and finally discharges into Lake Ontario at Rochester. It is one of the few rivers in the United States that flows from south to north. Forests, farms, and little villages line the river until it reaches Rochester, where great waterfalls powered nineteenth-century industries. Today, Rochester remains New York's third largest city, with more than 200,000 residents.

In 1972 the Weather Service Office at Rochester handled flood forecasting for the Genesee.[1] The Meteorologist-in-Charge retired not long before the Agnes floods, and no one had yet been tapped to head the office. Shorthanded and without supervision, the staff at Rochester faced its most crucial forecasts in a generation. Weather Service headquarters rushed in George Shienlein, a senior forecaster at Albany, to take the helm.[2]

Wednesday's forecast for western New York was "cloudy with a chance of showers". The showers started in Rochester the night before, but the city itself didn't receive much rain. Weather Service staff had no inkling of the deluge in the headwaters of the Genesee until 6:30 a.m. Wednesday morning, when a local radio station reported that schools at Wellsville, eighty miles south of Rochester, were closed because of flooding.[3] Rochester immediately issued a flash flood warning, but "it was probably too late".[4] Eight inches of rain fell at Wellsville by mid-day Wednesday,[5] raising the Genesee to a record stage of eighteen feet, flooding village streets and swamping riverside homes and businesses. Residents hoped the worst was behind them when the deluge diminished to a drizzle Wednesday afternoon. It was a false hope.

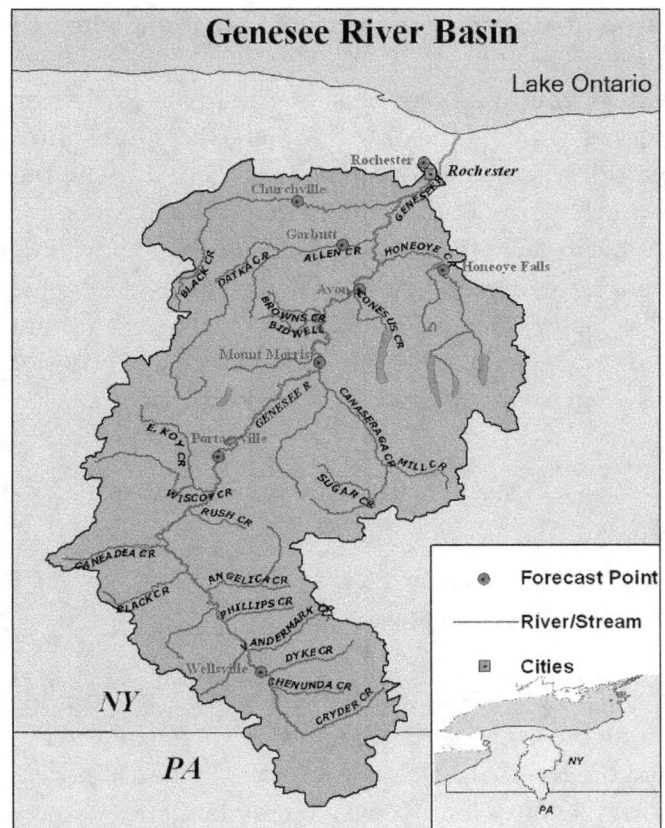

Genesee River Basin

Lake Ontario

The Genesee River flows north from the Twin Tiers to Lake Ontario at Rochester (NWS)

"We were close to the disaster stage even *before* Agnes got here," Shienlein told reporters.[6] When the core of Agnes unexpectedly veered westward and stalled over the Twin Tiers it brought a second blast of relentless rain, an additional six inches from Wednesday through Thursday. The Genesee River climbed four feet higher than ever before, with a discharge of 39,000 cfs – more than double the previous record.[7] At least two thousand residents of Wellsville – about a third of the population – evacuated to safety at schools and churches.

The continued heavy rain and subsequent flood devastated every riverside village on up through Allegany County. Every road in the county was closed to all but emergency vehicles,[8] and most of the bridges destroyed. At Scio, 21-year-old Joe Griffin, in too much of a hurry, drove around a barricade, evaded a flagman, and plunged to his death off a washed-out bridge.[9]

George Shienlein, acting Meterologist-in-Charge at Rochester, struggled to make sense of the little data that came in. "Only" three inches of fell around the city; he knew there was more upriver, but didn't know how much more. "Some of those guys who live near the river haven't been able to get to their gauges, and we don't know how much rain has fallen."[10] All three river gauges in the headwaters of the Genesee – Wellsville, Scio, Portageville – were knocked out.[11] Without precipitation and river stage data, the Weather Service at Rochester could barely guess what might be coming toward their city.

* * *

Thursday, June 22, 1972
Wellsville, New York

Hospital administrator Bill Foster looked nervously from his window as the roiling brown water of the Genesee River chewed big chunks of soil from the riverbank below his office. Jones Memorial, the only hospital in Allegany County, was in real danger.

Foster summoned his floor managers and gave the order: all patients and personnel in the west wing must be moved immediately. Nurses and orderlies worked past midnight to move fifty-five patients to the safer east wing, where many of them remained on gurneys in the hallways.[12] Custodial staff carried and rolled as much of the valuable medical equipment and records as they could out of harm's way. Finally, the chief of maintenance shut off the gas and electric lines to the west wing.

An hour later, Foster, and all of the hospital's staff and patients, held their breath, their hearts racing, as the hospital shook with a low rumble. The Genesee, at its record high, had undercut the very foundations of the building. With a shriek of twisting steel and crash of crumbling concrete the entire west wing of the hospital collapsed into the river. All the furniture, beds, and remaining medical

Jones Memorial Hospital and a nearby church collapsed into the raging Genesee River at Wellsville, New York. No one was injured (Dick Neal photo, courtesy of Allegany County Historical Society)

equipment broke apart and washed away.[13] Part of a nearby church and other riverside buildings suffered the same fate. Yet thanks to the quick work of Bill Foster and his hospital staff, there were no injuries. Jones Memorial Hospital remained open for the patients in residence and for new emergencies that came to its doors.

* * *

Friday, June 23, 1972
Mount Morris, New York

Forty river miles north of Wellsville, the Genesee River surged into Letchworth State Park, over the falls, and into the spectacular "Grand Canyon of the East". The ground shook like an earthquake as the swollen river crashed 350 feet down over the great cataracts.[14]

At the north end of the park stands the mighty Mount Morris Dam, built in 1952 to protect Rochester, fifty miles north, from flooding. At 230 feet high and a thousand feet wide, Mount Morris is one of the largest flood-control dams east of the Mississippi. When the river floods its waters can be held in a pool seventeen miles long, then slowly released to regulate the river's flow and reduce damage downstream.

Agnes was about to test Mount Morris Dam to its limits.

Sam Maiore, chief hydrologist at the Army Corps of Engineers' headquarters in Buffalo, monitored the swelling pool of flood water behind the dam. In constant contact with Shienlein at Rochester, civil defense authorities, and the dam superintendent at Mount Morris, Maiore had to decide if and when to release water from the reservoir.[15] The dam could be overtopped or even breached if a release was too little or too late, sending a deadly wave of water and debris into the villages and cities below.[16] On the other hand, if too much water was released, it would flood those same communities, although not as catastrophically as an uncontrolled overflow. Maiore had to release enough water to protect the dam while at the same time minimizing damages downstream, a precarious balance.

The reservoir filled quickly as flood water poured down the Genesee at the rate of 90,000 cfs – more than the average flow over Niagara Falls. Maiore opened the first spill gates at 4 p.m. Friday afternoon to relieve pressure on the dam, while local police evacuated residents and businesses down river in Monroe and Ontario counties.[17] By Saturday morning he opened all nine of the gates, spewing almost 30,000 cfs into the river towards Rochester, a hundred times the normal flow.

At Rochester, seven hundred National Guardsmen, joined by six hundred teenage volunteers, stacked sandbags along low-lying neighborhoods and infrastructure. Another 5,000 citizens offered to pitch in,[18] kept on stand-by but ultimately not needed. 2,300 homes were evacuated.

The controlled releases from the Mount Morris Dam worked. The Corps "deliberately flooded a 35-mile stretch of the Genesee

Valley and averted a major disaster."[19] The river peaked at 17.9 feet at Rochester, four feet above flood stage, on Saturday afternoon, the highest since construction of the dam. High Falls and Lower Falls thundered with the flood, and riverside roads swamped three feet deep. Yet the Genesee River caused little damage at Rochester. The Corps of Engineers estimated that if the river had not been controlled at Mount Morris, the full assault of 90,000+ cfs on Rochester would have been catastrophic, with damages totaling $210 million.[20]

LAKE OF SALVATION

June 21-24, 1972
Pittsburgh, Pennsylvania

In 1794 the revered Seneca chief Red Jacket joined President George Washington in the Treaty of Canandaigua, granting a wide swath of the New York - Pennsylvania borderlands to the Seneca nation forever. "Forever" lasted until 1960. Overcoming protests and lawsuits, the Army Corps of Engineers took ten thousand acres of Seneca land to build mammoth Kinzua dam on the Allegheny River near Warren, Pennsylvania. On the map the lake behind the dam is called Allegheny Reservoir. The Seneca people called it Lake of Betrayal.[1]

The Saint Patrick's Day flood of 1936 set records on the Allegheny River from its headwaters in upstate New York all the way to Pittsburgh. There, the river hit forty-six feet at Point State Park, twelve feet higher than ever before. Flood water inundated sixty-five percent of downtown. A hundred thousand buildings were destroyed, including many of the city's great steel mills. Forty-five residents of the Steel City lost their lives, and damages came to $108 million – more than $2 billion in 2020 dollars.[2]

Kinzua Dam was built to protect Pittsburgh and the Allegheny Valley from another Saint Patrick's Day flood. Completed in 1965, the dam is one of the largest in the eastern United States, providing flood control, hydroelectric power, and recreational boating, fishing and camping on the reservoir.

As with the Mount Morris Dam on the Genesee River, Agnes would test Kinzua Dam to its limits. And, as at Rochester, the dam would save the City of Pittsburgh.

The Ohio River Forecast Center at Cincinnati[3] was responsible for predictions of the Allegheny and other rivers in the Ohio River watershed. They had their own IBM 1130 computer, and their own algorithms and protocols for forecasting the Allegheny, Monongahela

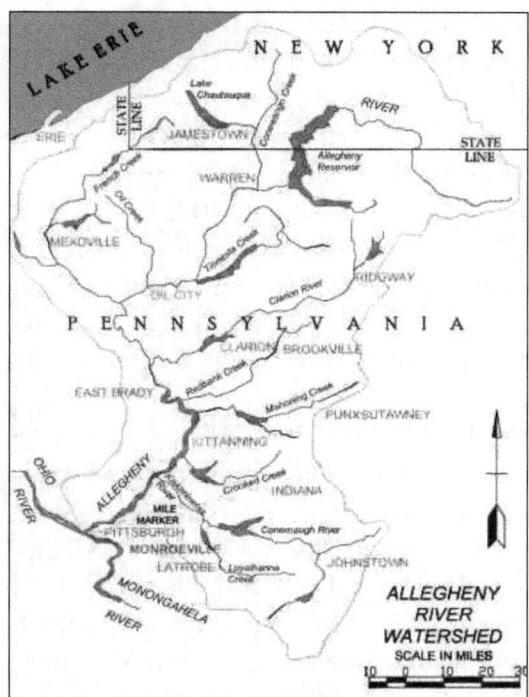

The Allegheny River flows from the Twin Tiers to Pittsburgh, where it joins the Monongahela River to form the Ohio River (NWS)

and Ohio rivers. Unlike the Susquehanna River monitored by the Middle Atlantic RFC, the Ohio River and many of its tributaries were controlled by dams and locks for transit of commercial and recreational watercraft. Coal mined in the Allegheny Valley a hundred miles east of Pittsburgh could be barged through locks down the Ohio and Mississippi Rivers all the way to New Orleans.

Hydrologist Bill Long compiled river stage and precipitation data at Pittsburgh's Weather Service office, then sent it to Cincinnati to calculate river forecasts. In turn, Cincinnati sent its forecasts back to Long for dissemination to the "customers" who needed them. Since flash floods on the upper Allegheny and its tributaries were a function of local rain and runoff, it was Long at Pittsburgh who issued flash flood watches and warnings for the communities upriver.[4]

The trouble began with heavy rain Tuesday night near the Twin Tiers village of Coudersport, Pennsylvania. The Allegheny River there is narrow, shallow and swift, winding northward into New York before bending back to towards Pittsburgh. Three and a half inches of rain fell at Olean, New York, overnight, bringing the river up swiftly. Volunteer observers alerted Bill Long to the downpour, and he issued a flash-flood warning for the upper Allegheny.[5]

The warning was heeded by Olean mayor Bill Smith and broadcast over WHDL radio, Olean's top-40 music station. By noon Wednesday the river rose nearly twenty feet, well above flood stage but not near the top of Olean's concrete dikes.[6] There was little damage beyond a few backed-up storm sewers. As predicted, the rain slowed to a drizzle Wednesday afternoon.

On Wednesday evening, June 22, the Ohio RFC ran its flood model for the Allegheny River, adding a hypothetical two inches of rain from Agnes – more than was actually predicted – in a "what if?" scenario. The model forecast no further flooding.[7] Already staying late, the hydrologist on duty at Cincinnati turned off the lights and went home. Agnes made her unexpected turn inland hours later, when there was no one at the Ohio RFC to hear about it.

The rain resumed and grew heavier around Olean as the core of Agnes neared; the Allegheny River resumed its climb up the city's flood wall, seven inches every hour. Mayor Smith called all police and fire personnel to duty, activated local civil defense volunteers, evacuated six thousand residents from their riverside homes, and called Governor Rockefeller for help from the National Guard. Staff at St. Bonaventure University, on the outskirts of town, hefted the library's collection of books and archives to safety on higher floors. For the next three days, Mayor Smith grabbed what sleep he could on a cot in his office as the flood emergency swirled around him. At WHDL, station manager Don McLean stayed on the air for eighty-nine straight hours, broadcasting updates to a panicked public.[8]

A total of nine inches of rain fell at Olean by Friday, with up to sixteen inches in the hills upriver. The Allegheny rose three feet higher than the previous record flood, inches below the top of the levee and spewing over in some low spots. It was so close that "if you mowed the grass on the dikes they would have overflowed."[9] The wall held, but damages in the little city ran in the millions.

The flood surged down the Allegheny from Olean to the village of Salamanca, a leasehold within lands of the Seneca Nation. With a flow of 73,000 cfs – besting the 1946 record of 49,000 cfs – the stage went nine feet higher than ever before, topping riverside dikes and all three of the bridges across the Allegheny.[10] Downtown Salamanca, including City Hall, the fire and police stations, and dozens of homes and businesses swamped in the flood. But residents heard and heeded Bill Long's flash flood warning. The flood devastated property in Salamanca, but no one was injured.

Downstream from Salamanca the Allegheny River entered the Allegheny Reservoir, stretching twenty-four miles to Kinzua Dam. Corps of Engineers reservoir manager Jack Ewers, at his headquarters in Pittsburgh, received teletyped precipitation and river reports from Bill Long at the Weather Service, data he needed to regulate flow from Kinzua and other dams in the Ohio River watershed. He began releasing water from Kinzua when rain from the core of Agnes moved inland Wednesday evening, making room in the reservoir for what was shaping up to be a flood of epic proportion.[11]

During the next three days Lake Kinzua rose thirty-four feet, within three feet of the crest of the dam.[12] Much like the Corps achieved on the Genesee River at Mount Morris, Ewers timed releases to minimize danger to the dam, while preventing overtop of the dam and flooding of villages down river. His careful management of the reservoir saved the village of Warren, just below the dam, from utter destruction.[13]

* * *

Kinzua and other dams and diversions throughout the Allegheny Valley held back less than a third of the watershed's flood waters.[14] Run-off of extreme rain from a multitude of western Pennsylvania tributaries ran unabated to combine in a massive surge down the Allegheny aimed directly at Pittsburgh.

At Pittsburgh, Bill Long juggled the changing forecasts and

updated data coming in from the Allegheny and Monongahela watersheds. He and his technicians stayed in constant contact with volunteer observers, the Ohio RFC at Cincinnati, the Corps of Engineers, Civil Defense authorities, the Pittsburgh Department of Public Works, city police and fire departments, and the media. The phones at the Weather Service couldn't handle the tidal wave of incoming calls, and many observers and other callers heard only a recording that the office was closed.[15]

Long barely slept in the three days since the rain started. Now, Thursday night, faced with a "weather combination that had never occurred before,"[16] he needed an updated river forecast immediately. But the Ohio RFC at Cincinnati again closed for the night, and no one was there to run the computer forecast models. Bleary-eyed, at 4 a.m. on Friday, Long plugged the latest river gauge and rain numbers into his charts and ran the algorithms by hand. The math told him that the river would surge over its banks at Pittsburgh by 8:30 a.m., sure to flood extensive low areas.

Long's flood forecast was made only hours before the morning rush hour. No sirens sounded, and media alerts came only with the morning news on radio and television. Civil Defense and Public Works staff relayed the warning to major riverside businesses,[17] but the flood caught most of the city by surprise.[18]

Point State Park flooded first, and its Fort Pitt Museum quickly filled with muddy water. Commuters driving to work downtown found bridges over the Allegheny, Monongahela and Ohio Rivers closed, the ramps leading to them submerged in the rising rivers. The flood poured into parking garages below ground level and destroyed the cars inside. Every office building and warehouse along the north side of the Allegheny River filled to its first-floor.[19] Commerce throughout Pittsburgh came to a halt.

Huge barges on the Allegheny River lifted from their moorings and washed into city streets. "Dozens of them floating in from the river, 150 feet long and ten feet high over the water. They were beached on flooded roads and stuck under bridges, spinning in the turbulence."[20] Other barges tumbled over the Highland Park dam and piled twenty feet high below.[21] Pleasure boats wrenched from their docks at posh "marina row" were smashed to bits when they collided with bridge piers or tumbled over dams at the river locks – a loss of more than $2 million.[22] The influential owner of the marina complained that he was not warned of the coming flood, and sparked an inquiry by the U.S. Congress.[23]

There were no deaths in Pittsburgh due to the Agnes flood, but there were close calls and daring rescues. Three longshoremen were swept off a breakaway barge, then treaded water in the Allegheny River before clinging to an abutment to the Ninth Street Bridge and plucked to safety by city police in a small boat.[24] A hundred police in the city were detailed to twelve-hour shifts to evacuate residents stranded in riverside homes to safety.

Three Rivers Stadium, home of the football Steelers and baseball Pirates, stood on the north bank of the Allegheny. The Pirates were out of town, and a huge concert was scheduled for Friday night: heavy metal star Alice Cooper was to rock forty thousand fans in the biggest concert Pittsburgh had ever seen.[25] Agnes forced a change of plans, flooding the field four inches deep and cutting off electric power.[26] The show was postponed until July, but it wasn't all bad – some lucky fans were stranded and "got to hang out that night with Alice and the boys" at the Golden Triangle Hilton.[27]

Fed by the flood surge from the headwaters, the rivers at Pittsburgh kept rising all day Friday and into the night, reaching an apex of 35.85 feet Saturday morning: eleven feet above flood stage, but ten feet below the record 1936 St. Patrick's Day flood.

The Ohio River raged from Pittsburgh down to New Cumberland and Wheeling, West Virginia, then on through eastern Ohio, twelve feet above flood stage, the sixth highest flood ever on the Ohio River. At least 10,000 residents evacuated islands and riverside villages, and one person died.[28]

On the north side of Ohio, near Cleveland, 55-mph winds in Agnes' counter-clockwise swirl whipped up fifteen-foot waves on Lake Erie that ripped into marinas and dockyards, causing more than a million dollars in damages.[29]

The Allegheny, Monongahela and Ohio Rivers at Pittsburgh began to recede on Saturday afternoon, and the massive cleanup could begin. After working around the clock for more than four days, hydrologist Bill Long could finally get some sleep.[30]

* * *

Above Kinzua Dam the Allegheny River far exceeded its previous record stage and discharge, bringing destruction to Coudersport, Olean, Salamanca, and other Twin Tiers villages. But the dam and other works saved the big cities below, sparing them from what would have been a disaster many times worse than it was. Kinzua may have been a Lake of Betrayal to the Seneca, but was a lake of salvation to the great city of Pittsburgh. The Agnes flood was the third highest in the city's history, and has not since been surpassed. Engineers reported that but for Kinzua and other dams, the flood would have risen to 47.9 feet, two feet higher than the 1936 St. Patrick's Day disaster.[31] Damages on account of Agnes at Pittsburgh amounted to "only" $45 million, but if the dams were not there could have exceeded $1 billion.[32] There is no way to know how many lives would have been lost.

THE FIREMAN

Wednesday, June 21, 1972
Gang Mills, New York

The Chemung River, a major tributary of the Susquehanna, begins at the confluence of the Tioga and Cohocton Rivers near the twin villages of Gang Mills and Painted Post, New York, just upriver from the little city of Corning. Six inches of rain fell over the region on Tuesday and Wednesday, rain produced yet not by the core of Agnes, but by the cold front sagging in from the Midwest. Fed by this downpour the Cohocton River at Campbell, New York, reached a height not seen since the record established there in 1935. On Wednesday afternoon Farley Stamp, an employee of Corning Glass Works, was swept from a stopped car into the raging waters of the Cohocton near the village of Bath, the first person in New York to die in the storm.[1]

All afternoon the Steuben County Fire and Police radios squawked with reports of flooding, rescues and evacuations along the Tioga and Cohocton rivers. At Lindley, on the Tioga River six miles upstream from Gang Mills, the flood was already higher than the previous record. That surge was coming directly to Gang Mills.

Hobart Abbey[2] parked his truck at the foot of the modest levee separating the Tioga River from the rail yard and nearby residential areas of the village. He had excused himself from his job at Corning Glass at noon, called to duty as assistant chief of the Forest View - Gang Mills Volunteer Fire Department. The levee, a long grassy mound ten feet high, had been built after the flood of 1946.[3] Abbey had to see if this flood could threaten his community.

As he scrambled up the grassy berm, Abbey heard the churning river even before he saw it. It sounded like a cement mixer, with its load of rocks and debris eroded from upstream rattling against the riverbanks. Reaching the top, in the fading summer daylight he saw

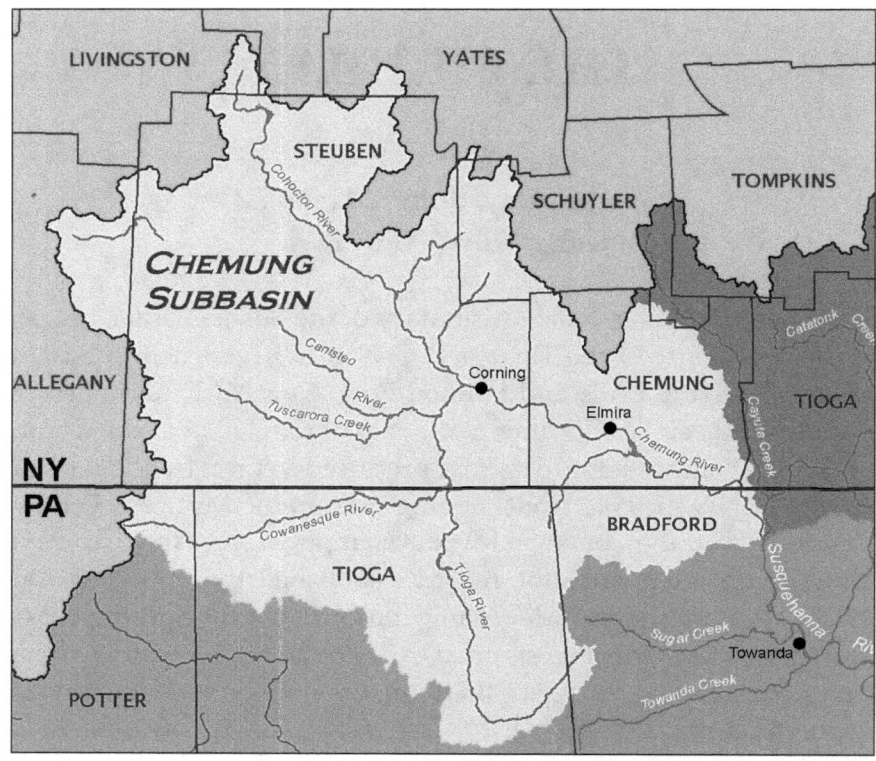

The Tioga and Cohocton Rivers meet just upstream from Corning, New York, to form the Chemung River, a major tributary to the Susquehanna (NWS, River Forecast Center)

the muddy torrent just two feet below where he stood. "Crap!" he muttered, as chunks of the earthen levee fell into the river, gouged away by the current. He raced back to the truck.

Fire chief Jack Kuehnle waited at their truck working the two-way radio. Built for rough terrain in forest fires, the "brush buggy" stood high on heavy-duty 30-inch tires, fully equipped with rescue equipment and the latest communications gear. As Abbey rushed toward him Kuehnle gave him an update. "County says they're evacuating Lawrenceville,[4] and wants us to assist." That village, with 550 residents, was only five miles upriver. Communities in Steuben County often shared first response duties, and Abbey served as county-wide training coordinator. Ordinarily, he would have marshaled the men and equipment of Gang Mills and rushed to the aid of Lawrenceville.

Abbey talked fast. "We can't do it Jack, we're in trouble. It's eating away the dike. If that levee goes ... God help us!"

They raced to the fire station on Forest Avenue where Abbey's 22-year-old daughter Betsy, a volunteer dispatcher for the department, blasted a radio call to all personnel, calling them to duty.

Abbey and Kuehnle worked south on Hamilton Street, parallel to the Tioga River. They knocked on the door of every home and urged residents to get to high ground. There were a few stragglers, but most people realized the danger, packed essentials, locked their doors, and went up the hill. Many found refuge in the homes of friends, relatives, and even strangers.[5]

Shortly before 10 p.m. a minimal levee gave way three miles upstream from Gang Mills. The sudden release swamped businesses, roads and railways as it swept down the west side of the Tioga Valley, on the inside of the levee at Gang Mills. When the surge reached the village it chewed at the earthen berm where Hobey Abbey stood two hours before. Ever larger chunks of soil and rock collapsed into the current, from both sides now. Within minutes a huge gap opened in the levee, and the full force of the Tioga River rolled into Gang Mills.

Abbey and Kuehnle stopped to radio other police and fire units when they reached the intersection of South Hamilton and Forest Drive. Suddenly a roar came from behind them. They turned to see a wave of frothing water blast through the break in the levee. The brush buggy tilted as the water rose above the tires, and the engine sputtered to a halt. Abbey sent an emergency call before the battery shorted out. "Our truck's stalled and the water's coming up fast! Is anyone close enough to help?" Betsy, on her radio at the firehouse, received her father's last message.[6] It was too late. Help couldn't reach the stranded firemen. The flood was too high, and too fast.

The cab filled with muddy water as the firemen scrambled into the bed of their truck. When that too submerged they climbed onto the roof. The rugged brush buggy shook violently from the raging current and debris in the water. No match for the flood, the truck lifted and rolled when water washed over the top. Abbey and Kuehnle lost their grip and floated away in the torrent.

The firemen treaded water as best they could. Kuehnle grabbed onto a phone pole along U.S. 15, and for the next five hours hung on for dear life. At dawn he saw that the current, though undiminished, flowed toward a house 100 yards away. He took a big breath, let go of his phone pole, and swam with the flow. When he reached the house he clambered onto the roof, to be rescued by other firemen with a boat five hours later.

Hobey Abbey wasn't as lucky. He was caught in a tangle of brush along the Tioga River, dragged under by the current, and drowned.

In 2015, more than forty years after Agnes, the Steuben County community established "First Responders Park" at the intersection of South Hamilton Street and Forest Drive, nearly the spot where firemen Kuehnle and Abbey were caught in the flood. A memorial in Abbey's honor is inscribed

Hobart A. Abbey, 1st Assistant Chief
Gave the supreme sacrifice during the flood of 1972
While in service with the Forest View Gang Mills Fire Department

THE MAYOR'S BROADCAST

Thursday, June 22, 1972
Corning, New York

The economy of Corning was built on the solid foundations of the famous Corning Glass Works. It was called the "Crystal City," renown worldwide for its industrial, scientific and artistic glass. But some areas had gone to seed. Market Street, once lined with thriving businesses, now hosted a shabby strip of bars, pawn shops and pizza joints. With urban renewal on their minds, the Chamber of Commerce and civic organizations were determined to rejuvenate the rundown Market Street corridor.[1]

Bisected by the Chemung River into "Northside" and "Southside", Corning had a long history with floods. As early as 1918 the city built dikes along the river, improving them in stages until complete in 1955.[2] The levees were designed to contain "100 year floods" like the one that hit the city in 1946, cresting at 37.7 feet on Corning's river gauge.

* * *

Newlywed Virginia Kricht worked as a bookkeeper at Corning Glass Works headquarters in Northside. On most days she had a great view from her window over the low-slung Corning Glass Center, home of the famous Museum of Glass, Steuben studios, and Hall of Science.

The veil of heavy rain this Thursday obscured Kricht's view. She could barely make out the Chemung River a couple of hundred yards away. It ran muddy and high, unusual for June, but how it looked during every spring thaw, nowhere near the top of the levees that ran along both sides. Traffic moved easily across the steel Bridge Street Bridge and the beautiful stone arches of Centerway Bridge.

Most residents were aware that the Tioga and Cohocton Rivers, tributaries to the Chemung way up at Hornell, Wellsboro, and Bath were flooding. Could trouble be coming to Corning too? Kricht's supervisor turned on a portable radio for everyone in the accounting office to hear. WCLI-AM, one of two radio stations in Corning, played top-40 hits punctuated by local news, weather and sports. In between plays of Elton John's "Rocket Man" and Neil Diamond's "Song Sung Blue" came reports of sewers backing up and water seeping into basements in nearby Gang Mills and Painted Post. The weather reports called for continued rain through Friday. No one was concerned the city itself would be flooded; after all, Corning's levees were high and stout, built to contain the river at its worst. But employees at Corning Glass and every other business did worry that the bridges might be closed. How would they get home from work?

Shortly after noon WCLI reported that workers at Ingersoll-Rand's big factory in Painted Post had been sent home because of rising water. Kricht's parents lived in Painted Post, and she rang them to make sure they were alright. The flood no longer seemed like idle chatter. Yes, her mother assured her, firemen came to their house to pump water out of the basement. Then she added, "but it filled right back in, and now it's higher than before." Kricht said she would try to get out of work early to bring her parents home with her for the night – just in case. She and her husband recently purchased a quaint Victorian home on Wallace Street in Northside, safe behind the city's levees.

Early in the afternoon, Amory Houghton, president of Corning Glass Works, sent employees who lived in Southside home early in case the bridges closed. The problem was that the bridge decks were six feet below the top of the levees, accessed via gaps that had to be plugged if the river rose high. Unlike at Wilkes-Barre, where big steel gates could be dropped into the bridge-ways, Public Works crews in Corning filled the gaps with sandbags and huge logs. They kept supplies at the ready, but it took several hours to get the job done. So far, no one at the city said the bridges *would* be closed, but if the river kept rising it *could* happen.

Kricht went to the window to again look over the river as word of the early dismissal spread through the office. The river didn't seem much higher than two hours before, but now cars streamed from the big Corning Glass factory in Southside and over the Bridge Street Bridge, headed to residential sections of Northside.

Then she noticed something odd. A line of rail cars filled with coal slowly backed onto the Penn Central Bridge from Southside. Coal trains ran across the river several times every day, but this was different. "For heaven's sake, what are they doing?"

One of her co-workers joined Kricht to watch as the fifteen heavy gondolas came to a halt on the bridge. The locomotive disconnected and crawled back towards the yard near the glass factory. "It looks like they are using the coal cars to hold down the bridge." It was a well-known practice to anchor a rail bridge with laden cars when threatened by rising waters.[3]

Kricht turned to her friend and laughed nervously, "Do you think they know something we don't?"

* * *

Thursday, June 22, 1972
Harrisburg, Pennsylvania

O.D. White and his hydrologists at the River Forecast Center received scant data from the distant reaches of the Susquehanna River's tributaries. Stations in the Twin Tiers still able to report were few and far between, with none in the upper headwaters.

Andrea Miller, a veteran observer at Addison, New York, reported four inches of rain and a record crest on the Canisteo River, a tributary to the Chemung River, on Wednesday,[4] but her report didn't capture the surge coming from much heavier rain upriver. It would be the last report Miller made: by that night her gauge and instruments flooded over, and her own home was inundated. Not only at Addison, but gauges on the Canisteo River at Hornell, on the Tioga River at Lindley, and on the Cohocton River at Avoca washed out.[5] At Corning, the volunteer observer quit months before, and White had not yet found a replacement.

The hydrologists couldn't speculate about what might be happening where didn't have new data, but could use only the facts they had. If a gauge flooded over, they used the stage at which the gauge stopped working; when precipitation reports ceased, they used only the reports from the few stations still able to check in. These spotty measurements didn't capture the much heavier rains, and flood waters rising, higher upstream. An observer of the Chemung River near Elmira reported "only" seven inches of rain, but it was later determined that up to sixteen inches fell in the far headwaters, an unseen deluge that roared toward the unsuspecting residents of Corning.

In 1972 the RFC did not prepare river forecasts for Corning but did for Elmira, eighteen miles downriver. Hydrologist Nick Pavick sent an updated report to the Weather Service Office at Binghamton late Thursday afternoon, predicting that the Chemung River would crest at Elmira at 11 p.m. Thursday night, then begin to recede.

* * *

Thursday, June 22, 1972
Corning, New York

Corning's Public Works superintendent Chuck Moynihan called Al DeRenzo, his counterpart at Elmira, every hour to get the updated river forecast. Moynihan had a chart, penciled on graph paper, to translate the forecast for Elmira into an estimated stage at Corning. It was a "rule of thumb", not exact, but usually sufficed. When Moynihan received the afternoon forecast he calculated that the river would peak nine feet below the top of the Corning's levees later that evening. He would have to plug the gaps at the bridges, and worried about flood waters infiltrating the city's water supply wells, but this forecast put Corning in no real danger.

Corning's evening newspaper, *The Leader*, hit the streets at 5 p.m.

An Associated Press story on the first page told about flooding in Richmond, Washington, and Harrisburg. These cities were hundreds of miles away, of little concern in upstate New York. The second page had short articles about sewers backing up in nearby Painted Post, and of a man thought to be drowned in the Cohocton River at Bath. A story on the third page was headlined:

STURDY DIKES SAVE CORNING ONCE MORE

Despite a few streets awash and basements flooded, *the dikes are holding and there is no danger for city residents.*[6]

It would be two months before *The Leader* could again publish in Corning, after its offices and plant were utterly destroyed in the coming disaster.

* * *

Glass Works president Houghton released the rest of his employees at 4:00 p.m. Ginny Kricht was getting ready to go home when her mother called. "You don't have to come get us dear. We are going up to the school for the night." Kricht had heard WCLI's report that the Red Cross opened an evacuation center at Painted Post High School. Her mother continued before Kricht could argue. "A fireman came to the house and told us they want everyone along Chemung Street to leave for the night. Just in case, you know. And dad wants to stay in town to help if they need him." Kricht's father was a retired physician.

Kricht protested weakly, but was satisfied her parents would be safe. She took the elevator to the ground floor, got drenched by the rain as she scurried to her car in the Glass Center's big parking lot, and drove the few blocks to her home.

Still fretting about her parents, Kricht kept the radio on as she made supper. The local news was all about the floods: residents in low-lying areas of Gang Mills and Painted Post had evacuated to the high school; police were still looking for the man lost in the Cohocton; some of the roads along the rivers were closed. And then,

"Mayor Joseph Nasser of Corning has recorded a statement about the flood situation."

> *I want to thank the people of Corning for their patience and understanding in this emergency. Please be assured there is no danger in Corning. Our Department of Public Works predicts that the river will not rise higher than nine feet below our levees sometime this evening. As a precaution, I have directed the Department of Public Works to close the bridges across the river, effective immediately and until the river goes down.*[7]

Sighing with relief at the mayor's assurance, Kricht switched off the radio. After supper she settled down with her husband to watch *The Waltons*, her favorite TV show. Like most of the residents of Corning, the Krichts went to bed believing they were safe behind the city's levees.

Mayor Nasser's broadcast was based on the River Forecast Center's afternoon prediction, the best information he had at the time. He had no inkling of the incredible rainfall supercharging the headwaters of the river that bisected his city, nor of the veritable wall of water coming his way.

* * *

At 9 p.m., John Fox,[8] director of the Corning Glass Center, called Public Works superintendent Moynihan for an update. Like everyone else, Fox knew the Chemung River was rising, and although he heard the "safe" forecast from Moynihan personally, he was still nervous. Fox was responsible for security of the Glass Center, and didn't have enough maintenance staff to protect it flood water came into the city.

Fox checked the doors to the Glass Center one more time before going to his home a few blocks away. Along the way, he saw public works trucks delivering big logs and sand to the bridges, and police setting up barricades along riverside streets. If the city was taking precautions, he decided he would too, and moved his family from their low-lying home to his office at the Glass Center. Some neighbors and out-of town guests were visiting, and he invited them to come along. At 11 p.m., John Fox and his family, plus their neighbors – in all, seventeen people and two dogs – carrying sleeping bags and a change of clothes, rode the elevator to the top floor of the Corning Glass Center to spend the night. They had a birds-eye view of the twinkling city around them, bisected by the dark strand of the churning Chemung River.

* * *

Just before midnight, Mayor Nasser, superintendent Moynihan, the fire and police chiefs, the city health officer, and a liaison for the National Guard huddled in an impromptu "war room" at City Hall. The river was higher than the afternoon forecast, and rising a foot every hour. What was coming, and what could they do about it? Without an update from the Weather Service at Binghamton and DeRenzo at Elmira, they had to prepare for the worst. The mayor called Governor Rockefeller for permission to activate the local National Guard, and directed police to patrol the dikes for leaks.[9] Moynihan grabbed any volunteers he could find to speed the crucial job of closing the bridge gaps in the levee.

* * *

Seventeen-year-old Larry Elliott, with his friends Bob Bush and Tim Fraser, volunteered to fill and stack sandbags at the Bridge Street Bridge. Public Works employees used front-end loaders to hoist phone poles horizontally into the gap, while other workers and volunteers filled canvas bags from a pile of sand. They packed the bags around the logs to hold them in place, then piled more on top. The makeshift barrier rose through the night, illuminated by glaring portable floodlights.

Soon after midnight, with the sandbagging job done and the gap to the bridge filled, Elliot and his friends, still wide awake with excitement, walked the top of the levee downstream toward the

Penn Central railroad bridge. The swollen river, choked with debris, surged several feet below the levee top. The boys weren't alone, as a crowd of onlookers assembled on the levee and across the bridges, mesmerized by the violence of the roaring Chemung River[10] The boys speculated what it might be like to ride the rapids in a canoe – suicide, they concluded.

The gap in the dike at the railroad bridge had been barricaded in the same way as Public Works had done at Bridge Street. The boys gawked at the load of flotsam jammed against the upstream side of the bridge, partially damming the river. In a few places the river sloshed over the railway on the bridge and splashed up against the laden coal cars parked upon it.

* * *

At City Hall Mayor Nasser, Moynihan and the others stayed on watch. When the river stopped rising at 1:00 a.m., they thought they had dodged a bullet, and some of the staff went home to bed.

At 3:00 a.m. DeRenzo, at Elmira, received a new forecast from Binghamton, and called Moynihan: "Now they say the river will crest at twenty-four feet here in Elmira at 7:00 this morning. You're two or three hours ahead of us, so figure 5:00 at Corning."

Moynihan eyes widened as he ran his finger along his chart. This new forecast said the river at Elmira would reach twenty-four feet, but his chart didn't go that high. It showed a maximum of only twenty-one feet, the 1946 record. If this forecast was true, the Chemung River would spill over the top of Corning's levees. DeRenzo wryly added, "We don't believe it either."[11]

Believing it or not, Mayor Nasser prepared for the worst, and directed the police, fire department, and National Guard to get residents out of low-lying areas. But in the middle of the night almost everyone was asleep, and there were too many homes to knock on every door.

Moynihan recorded a statement played over and over on WCLI, warning residents in riverside areas of Corning, both Northside and

Southside, to evacuate to schools on high ground. *We do expect that water may spill over the dike.*[12] Few people were awake to hear it.

The surge from the unseen deluge in the headwaters of the Chemung River hit Corning at 5:00 a.m. It overwhelmed a dike at Painted Post, and sent a wave of water rushing into Northside on the *inside* of the levee.

The destruction of the Crystal City had begun.

SHATTERED GLASS

Friday, June 23, 1972
Corning, New York

It was well after midnight before the Fox children settled into their sleeping bags in the executive offices at the Corning Glass Center. Their parents peered out the rain-streaked windows with a mix of fascination and dread. Public Works crews sandbagged the gaps in the levee, debris in the river piled up against the coal-laden Penn Central Bridge, and emergency vehicles darted through the city. They heard Public Works superintendent Moynihan's statement for a third time. Was it really possible the river could come over the dikes?

When police reported the 5 a.m. break in the levee at Painted Post, Mayor Nasser knew there no hope of avoiding the flood, and ordered the fire chief to blast his sirens throughout the city. The sirens were soon joined by the whoop-whoop of police cars and the honking of city trucks. Residents drove along the streets, blaring their horns to alert their neighbors. Above them all shrieked the great whistle of Corning Glass Works,[1] an unceasing wail to raise hairs on the back of every neck. Get up! *Get out!*

John Fox, with his family and neighbors at their refuge in the Glass Center, gaped in disbelief as the wave of flood water from the break at Painted Post rushed through the streets of Corning's Northside, sweeping a tangle of debris ahead of it. Just as the wave reached the Glass Center someone shouted. "My God, the bridge is falling!" Eyes turned to the Penn Central railroad bridge as its massive steel girders twisted, shifted off their piers and collapsed into the churning Chemung River. The heavy gondolas parked on the bridge to anchor it pitched sideways and spilled their burden of coal.

The fall of the Penn Central Bridge freed the dam of flotsam caught on its upstream side. Flood water released from the broken

Corning's Penn Central railroad bridge, burdened with railcars to weigh it down, collapsed into the raging Chemung River. Corning Glass Works headquarters and the Corning Glass Center are beyond the levee in the background (Courtesy of Corning-Painted Post Historical Society)

dam came like a tsunami, some said ten feet high, to slam into the levee at the Center Way Bridge and explode through the sandbag barrier only yards from the Glass Center. Cars that Fox and his neighbors parked high on the ramp to the Center's garage were swept away like toys. The wave imploded the big glass doors on the first floor and quickly filled the lobby. Fox and his wife clung to their children as the Glass Center shook violently, afraid the building would collapse.

Houses on Fox's own street, across Center Way, were ripped from their foundations and slammed by the torrent into adjacent homes. Just an hour before some of Fox's neighbors slept in those houses, but he could not know if they escaped. Some did not. The lights in the Glass Center flickered and died. All of Northside went dark.

Rising flood water thundered by the Glass Center, carrying with it debris that smashed the huge plate-glass windows on the first floor. The executive office on the second floor was no longer a safe refuge, so Fox guided his family and guests up the metal stairs to the rooftop. The young children whimpered as they clung to their parents, and huddled under carpets ripped from the floors below in a futile attempt to ward off the pounding rain. They would wait on the roof for five hours until rescued by helicopter.[2]

From below and inside the Glass Center, on treble notes above the bass rumble of the flood, glass shattered as display cases in the Museum of Glass and the Steuben studio toppled in the swirling water. Cabinets lifted and crashed into the walls, splintered into piles of pick-up sticks. Priceless glass art and artifacts rained from the shattered cabinets and displays, swallowed in the muddy maelstrom. Fourteenth-century illuminated manuscripts floated to the ceiling in the basement library. The world's greatest repository of glass history and scholarship, thousands of volumes of leather, parchment, and hand-made paper, saturated with oily, muddy, flood water.[3]

* * *

The sirens shrieked minutes after exhausted teenage sandbagger Larry Elliot slipped into his bed. Roused again, Elliot phoned Bush and Frazier, the friends who help with the sandbags. "Did you hear? The river's coming over the dikes!"

The boys returned to Bridge Street, where they piled sandbags only two hours before, to find the Public Works crew pulling away. The driver leaned out from his truck to yell over the drum of the rain. "Nothing more we can do here. We're going to get the boats. You get your families up to the high school. Go! Good luck!"

Elliot and his friends scrambled to the top of the levee to see it for themselves. As they gaped at the river swirling now just inches below, a screech of metal pierced the roar of the flood. They turned to see the trusses of the Penn Central Bridge twist and collapse a hundred yards from where they stood. It seemed unreal. "Let's get out of here!" It was 5:30 a.m.

The boys ran the few blocks to Elliot's house on Watkins Road. "My dad has a rowboat. We can help, like that guy said." It was not unusual for people in Corning to own small boats for fishing on the river or one of the nearby lakes.

The wave through the broken levee at Painted Post surged from the west, while backwash from the break at the Centerway Bridge gushed in from the east. The boys heard, then saw, it coming. They dragged the boat trailer to Elliot's driveway, released the little craft into the rising water, and clambered aboard.

* * *

A half hour later, 6:00 a.m., the river poured over the dikes into Southside. Like Hercules' purge of the Augean stables, the waters of the Chemung raged fifteen feet deep on Market Street to gut the bars, strip clubs and pawn shops. Residents on the second and third stories above the businesses waved sheets and towels to attract rescuers. Urban renewal came to Corning's seedy Southside in liquid form.

The river swept into Southside's big Corning Glass Works plant, snuffed out the great furnaces, and wrecked the precision manufacturing equipment. The glass workers had been evacuated, but the flood wreaked tens of millions of dollars in damages.

* * *

Over the next two hours teenagers Elliott, Bush and Fraser rescued twenty people from homes along the streets of Northside, and discharged them on higher ground. The little rowboat foundered as the water came faster and deeper, but Bush's family had a bigger boat with a motor. They commandeered it to patrol up and down the residential streets along Pulteney Avenue, picking up two or three people at a time from second story windows.

Police in a helicopter overhead shouted down with a bullhorn to direct the boys to people needing rescue. "Wallace Street, three houses down, two on a porch roof!" Elliot saw a man and woman

The Chemung River roared down Corning's Market Street, gutting the bars, pawn shops and strip clubs (Courtesy of Steuben County Historical Society)

frantically waving a white sheet. The turbulent current was swift, perpendicular to their course, and the boys gingerly guided the boat to the end of the porch. The woman – it was Ginny Kricht – handed a bawling cat to Elliot, and Frazier extended his hands to help Kricht and her husband aboard.

The crackle of splintering wood came from inside the house as flood waters ripped at the Kricht's home. No sooner had the group motored away when the trunk of a huge tree, torn from upriver and propelled by the flood, ram-rodded the venerable old home and smashed it into kindling. "I watched my house, that I had escaped from just minutes before, be swept away like dust from a broom,"[4] Ginny later told a reporter.

By the time it ended, Larry Elliott, Bob Bush and Tim Fraser rescued sixty people from the flood waters of Corning. Sometime after noon, awake for thirty hours, they retreated to the Bush home, above the flood, and collapsed to sleep until the next morning.

On Sunday afternoon, after the water went down, teenagers Larry Elliott and Bob Bush walked to the neighborhoods where they saved so many lives. Residents, some of the very people they rescued, scowled at them from the wreckage of their homes. "These long-haired kids shouldn't just be hanging around when they could be doing something to help!"[5]

* * *

Twelve hours after The Leader published *Sturdy Dikes Save Corning Once Again*, the Chemung River slammed into the newspaper's offices and plant on Pulteney Street, eight and a half feet deep. The newspaper salvaged its two big presses, but everything else was a total loss. No furniture, phones, typewriters, paper, not even pencils remained, all swept away or buried in mud. Most of the reporters, editors, and staff had evacuated, and, with the phones down, were incommunicado.[6] As at the Harrisburg Patriot-News, there was no way the Leader could publish on Friday.

Ed Underhill, publisher and editor of The Leader, met with Mayor Nasser and other officials at the City Hall war room late Friday evening. Without phones, without radio or TV, without power, the people of Corning had no way of knowing what was happening, what to do, or where to go. When can we go home? Can we drink the water? Why is the Army here, with guns? Where is my wife, my husband, my child? Rumors began to circulate: the entire city is being evacuated, looters are everywhere, poisonous snakes have invaded![7] The officials agreed: The Leader *must* publish *something* to convey essential information and allay the city's fears.

By word of mouth, Underhill gathered five of his seventy staff at Corning's junior high school, where they typed out a three-page Special Edition of The Leader. They made 1,500 copies on a hand-cranked mimeograph machine, and carried them by hand to the evacuation shelters. It looked more like a sixth-grade classroom newsletter than a commercial publication, but was deadly serious.

The three-page Leader, passed from neighbor to neighbor, offered the only source of official information in Corning in the first days after the flood. The entire first page consisted of a bullet-point statement from Mayor Nasser: *The water has gone down ... there is no mass evacuation ... the National Guard is here to protect people and property ... there is food, water and medical services at the evacuation centers ... the Red Cross is on its way ... do NOT come to the downtown areas.*[8]

The Leader rebuilt from scratch, publishing from the print shop of The Tribune in Hornell, New York, until its own refurbished offices were up and running again in August.[9]

* * *

John Fox slogged through the mud to the front doors of the Museum of Glass when the flood receded Saturday morning. A pair of armed National Guardsmen met him, stationed there to keep gawkers, or worse, looters, away. Fox blanched when he saw the big doors broken and ripped from their hinges. He didn't dare imagine the horror he would find inside.

When Fox stepped through the shattered doors the nightmare came true. The flood reached fifteen feet deep in the museum, and the collections were utterly devastated. Many of the display cases had toppled, spilling and shattering the antiquities inside. Muddy flood water infiltrated the cabinets that didn't fall, stirring and smashing the artifacts they contained. What days before were elegant displays of glass art were now piles of mud-crusted rubble. In the library, ancient manuscripts and archives were submerged for hours, their bindings swollen and disintegrating, their pages soaked and fragile, ripe for mold and mildew. A foot of oily, stinking mud covered everything. It was the worst catastrophe ever suffered by an American museum.[10]

Museum president Tom Buechner was at a conference in Mexico. Science director Bob Brill was in Afghanistan to visit an ancient glass works. The power and telephones were out, and virtually all roads were closed. Somehow, Fox had to get the museum management team back as soon possible.

* * *

Sunday, June 25, 1972
Kabul, Afghanistan

It was Saturday evening in Corning, but 6 a.m. Sunday in Kabul. Museum of Glass science director Bob Brill awoke to the persistent ring of his hotel room phone. He answered groggily, still not completely over the jet lag of his travels. The caller was from the U.S. Embassy. "Sir, we have a telegram for you. Would you like me to read it?"

Brill stirred. "Yes, of course." What could it be at this hour?

"From John Fox, to Dr. Robert Brill." The embassy man paused.

"Well? What does it say?"

Museum destroyed. Return Immediately.

Just four words. Brill was instantly awake, his stomach gone hollow. He had no idea what could have happened. It occurred to him that a plane crashed into his museum. It would be two days before Brill could reach anyone back home to hear the full story, and four more days before he arranged his return.[11]

* * *

At its peak, the Chemung River at Corning swelled to a discharge of 228,000 cfs, more than double the record of 1946. The flood poured over and the through the dikes for eighteen hours, destroying homes, businesses, schools and churches. The very asphalt of the city's streets ripped away, leaving only ragged gullies where cars and buses once ran. Six thousand people – a quarter of the city – evacuated, most of them returning to find their homes terribly damaged, often destroyed. Electricity, telephones, gas, water and sewers were out for weeks. The great factories of Corning Glass

Works and Ingersoll-Rand were ruined, and it was unclear if they could re-open. Damages were pegged at more than $172 million ($1 billion in 2020 dollars).[12]

No community suffered more human loss than Steuben County: twenty-four were dead, including eighteen in Corning and Painted Post. Residents went to bed assured they were safe behind the city's levees, and few were awake at 3 a.m. to hear the urgent pleas to evacuate. By the time the sirens sounded at 5 a.m. it was too late. There were not many deaths of people attempting to drive through flooded streets, as happened in other areas. Rather, in Corning, the victims were found in their beds, their basements, their shops, and their back yards.[13] As The Leader headline a week later said, *No One Knew the Flood Would Hit.*[14]

DERENZO'S PLAN

Thursday-Friday, June 22-23, 1972
Elmira, New York

In 1870 Samuel Clemens, better known as Mark Twain, married Elmira girl Olivia Langdon. For twenty years the family summered at Quarry Farm just outside the city. It was here overlooking the placid valley of the Chemung River that Clemens wrote his greatest novels, including *Tom Sawyer* and *Huckleberry Finn*.

Elmira, the "Queen City of the Southern Tier", could be called Corning's older, bigger sister. Only eighteen miles downriver from Corning, Elmira is likewise divided by the Chemung River into "Northside" and "Southside," knit together by several bridge crossings. In 1972, Elmira had 40,000 residents, more than double the population of Corning, with vibrant neighborhoods of Italian, Irish, German and African American heritage. The city had long been a hub of regional transportation and industry.

During the third week of June, 1972, the Chemung River wrought unfathomable destruction to the Queen City. The Agnes flood at Elmira was economically more devastating than at Corning, but in other ways the flood's impact upon Elmira was dramatically different.[1]

Unlike Corning, which was anchored by a single bedrock industry, Elmira's economy was more diverse. Manufacturers such as American Bridge, General Electric, Westinghouse and Remington Rand had big plants in Elmira, but their headquarters were out of state. By 1972 some of the factories had down-sized or even closed. The opening of a shiny new shopping mall in nearby Big Flats drew traffic away from the mom-and-pop shops on Water Street, and more than a few of the old storefronts shuttered. The population began to decline.

The headlines of the Elmira Star Gazette that Thursday, June 22, trumpeted Sen. George McGovern's victory in New York's Democratic primary election, sewing up his party's nomination to run for president against incumbent Richard Nixon. There was news of the Watergate scandal: one of the burglars, McCord, had ties to the White House. B-52 bombers continued to pummel North Vietnam. The weather blurb called for "a chance of showers."

A short article below the fold told of flooding on distant tributaries of the Chemung River, way up at Hornell. As at Corning, there was little concern about flooding upriver; after all, Elmira's flood walls, twenty-three feet high, were built to hold back a flood at least at bad as the 1946 "big one". That historic flood crested at 21.5 feet on Elmira's river gauge, bringing 132,000 cubic feet of water per second,[2] far more than ever before. The Army Corps of Engineers said such a flood was so rare as to be statistically insignificant, something that might happen once in 700, even 1,000 years.[3]

* * *

At his People's Place shop on Elmira's Water Street, Tommy Hilfiger, twenty-one years old, sold bell-bottomed jeans and peasant shirts, along with incense, black-light posters, and other accouterments of youth culture. The store was a thriving success. "Everybody wanted to wear clothes from the People's Place. It was the place to go for anything cool."[4] But this rainy Thursday Hilfiger's business was so slow he closed his store early, a few minutes before 5 p.m.

Hilfiger tuned his car radio to WENY, Elmira's most popular radio station. Sammy Davis Jr.'s "Candy Man" and Dr. Hook's "Sylvia's Mother" led into a report of local news: flooding up at Hornell, the apparent drowning at Bath, a few riverside roads closed. WENY's weather report said the river would rise at Elmira, but remain far below the city's flood walls. Hilfiger wanted to see for himself, and drove up Harris Hill to look over the whole valley. The City below was shrouded with rain, but Hilfiger could make out the spread of the Chemung's brown waters over farmland north and south. He realized that if the river swelled much more and rose into downtown Elmira, his stock at Peoples Place would be ruined.

Hilfiger stopped at his apartment to call his partners and his parents, then sped downtown. He was surprised no one was working at the shops around his. Over the next seven hours Hilfiger, his mother and father, and two partners carried all of Peoples' Place's goods from the basement storage and shop area up to the second floor.[5] They went home, exhausted, at 3 a.m. on Friday – just before the sirens began to wail.

* * *

Al DeRenzo, Elmira's director of Public Works, had prepared a detailed flood response plan in 1969.[6] Now, with an eye on the pummeling rain and rising Chemung River, he studied it all day. Flooding, according to DeRenzo's plan, would begin along Newtown Creek in Northside and Seeleys Creek in Southside. Storm sewers would back up, and some river water would infiltrate under and through the flood wall.

The plan called for three steps if danger threatened, progressively more severe: (1) a general alert and preparations, (2) evacuation of low-lying areas and closure of some roads and bridges, and (3) more widespread evacuations and mobilization of the National Guard. The city's fire, police, Public Works, and county Civil Defense rehearsed the plan just two months before. They knew what to expect – to a point – and knew what to do.

Elmira Mayor Richard Loll convened a flood task force at 8 p.m. Thursday evening, as called for in the plan. DeRenzo, City Manager Joe Sartori, the police and fire chiefs, and expert staff met at their command post in City Hall, a beautiful colonnaded edifice two blocks from the river, and prepared to take the actions spelled out in DeRenzo's plan.

DeRenzo told the task force that the Chemung River was at fifteen feet, eight feet below the top of the concrete flood walls protecting downtown Elmira. The latest forecast from the River

Forecast Center, via Binghamton, predicted a crest of seventeen feet at midnight. DeRenzo didn't think the Chemung would cause any harm, but there would be flash flooding on the little tributaries.

At first, problems arose just as DeRenzo predicted. Newtown and Seeley Creeks, fed by seven inches of local rain, overflowed into Elmira's industrial complex. Storm sewers backed up into some intersections, and a used car lot was swamped. Police threw barricades up on creek-side streets.

The rain continued, and the river kept rising. At 9:45 p.m., Thursday, City Manager Sartori ordered implementation of Step One. He went on local radio and television to issue a general alert: *Remain calm. Gather your important papers, medicines and some clothes. Be prepared to evacuate.* Police closed the Walnut Street Bridge to all but emergency vehicles. Public Works crews sandbagged low areas near the water treatment and sewer plants, and at vulnerable spots along the dikes. Forty-two people, two dogs and a cat evacuated from a flood-prone area in Southside to the Twin Towers dormitories at Elmira College,[7] the first of what would be half the population of Elmira moved out of harm's way.

At 10:30 the task force met again at its City Hall command post. DeRenzo reported that the Chemung was at eighteen feet, already higher than the predicted crest, steadily rising two feet every hour. He still didn't have an updated forecast from the Weather Office at Binghamton.

Police, fire and Public Works radios squawked in the command post with reports of flooding along the creeks in ever-widening areas of Elmira. Phones rang incessantly as residents and businessmen wanted to know if their homes, shops and factories were in danger. The task force didn't want people to panic; should they go to Step Two of the plan, close more bridges and roads, evacuate more residents? Without a new forecast, and with the river rising much faster than expected, the task force reached a consensus: skip Step Two and go directly to the ultimate Step Three. Mayor Noll declared a state of emergency and prepared for widespread flooding. He directed police, fire and Public Works personnel to evacuate the residents of Southside and the low areas of Northside. Governor Rockefeller authorized mobilization of the National Guard at Elmira, and the Red Cross set up shelters at city schools on high ground.

The turning point came at 2:30 a.m. At last, DeRenzo received an updated river forecast. It said the Chemung River would crest at twenty-four feet, at 7 a.m. – a foot higher than the city's 23-foot dikes. If this forecast was true all of downtown Elmira – the businesses, industries, homes, schools, churches, hospitals – would be inundated. It was hardly believable. DeRenzo's plan didn't imagine the river spilling over the walls.

Fifteen thousand residents had to be evacuated in four short hours. The task force didn't have the six hours anticipated in DeRenzo's plan. Police and firemen rushed a canvas of Elmira's neighborhoods, up one street and down the next, bullhorns blaring, *Evacuate your homes ... Leave immediately!*[8] The city's three radio stations broadcast the warning to Elmira's sleepy residents over and over: shut off water, power, and gas; bring important papers and medicines; alert your neighbors; go to the shelters, right now! Residents phoned family, friends, and neighbors, waking them and spreading the alarm.

Only a few at first, but soon a steady line of tail-lights traced up Elmira's avenues to refuge on high ground. Even in these wee hours, the people of Elmira were on the move.

Two hours later, at 5 a.m., with daylight brightening the drizzly skies, DeRenzo received a radio message from Chuck Moynihan, superintendant of Public Works at Corning. The Chemung River had topped Corning's walls and was ravaging the Crystal City. Power and phones were out, bridges collapsed, several deaths confirmed. A catastrophe beyond their worst fears had hit Corning eighteen miles upriver. DeRenzo knew it would slam into Elmira in two short hours.

Mayor Noll, city manager Sartori, and DeRenzo urgently picked up the pace of the evacuation. City and state police, firemen, volunteers, and now the first wave of National Guardsmen pounded on every door half a mile from the river on both sides. DeRenzo dispatched city trucks, transit buses and taxis to take people from apartment buildings downtown.[9] All but certain the flood would

The Chemung River poured over Elmira's downtown dikes "like Niagara Falls" (Bud Longwell, courtesy of Julianne Longwell Combs)

exceed the most dire conditions in DeRenzo's plan, the task force had to ad-lib. Still, there was little panic. Many residents even become sightseers, mesmerized by the rising river and the imminent destruction of their city.

The river reached twenty-three feet at 6:30 a.m., the very top of the gauge and the wall. Within minutes both would submerge in the flood.[10] Water already sloshed over low spots and seeped through cracks in the wall. DeRenzo pointed out that when the river came over the dikes – it was no longer a matter of "if" – all of downtown Elmira, including the command post at City Hall, would flood. With help from city maintenance staff, the task force gathered all the maps, plans, documents and equipment they could carry, and in a short convoy of city trucks evacuated to the administration building at Elmira College, on high ground, to set up a new command post.

By 7 a.m., the count of evacuees at the Twin Towers of Elmira College rose to two thousand. Other refugees streamed into schools, churches and public buildings in both Northside and Southside. They carried sleeping bags, suitcases, cardboard boxes and plastic leaf bags bursting with clothing and personal items. Some agitated evacuees shouted and shoved where they queued up to sign in at the shelters, while dogs strained at their leashes and barked at the chaos around them.

DeRenzo drove his Public Works truck to the Walnut Street Bridge, where his crew still piled sandbags. Further sandbagging was futile, and DeRenzo would assign this crew to rescue duty. When he opened the door of his truck he staggered at the acrid stench overpowering the earthy odor of the flooding river. Gasoline! Two storage tanks upriver at Big Flats had ruptured in the flood and spilled a half million gallons of high-octane fuel, coating the river at Elmira with an oily rainbow sheen.[11] Some of DeRenzo's men coughed and rubbed their eyes at the noxious fumes.

As the crew clambered into their trucks, two fully intact houses – from who-knows-where upstream – bobbed down the river and crashed into the bridge.[12] The collision cleaved one of the houses in half, leaving its roof strewn on the bridge deck while the lower half splintered apart. A short time later the entire bridge, a main link between North and Southside, collapsed into the oily, swirling Chemung River. No one was there to see it happen.

The river came over Elmira's dikes "like Niagara Falls".[13] White-water rapids raged through the downtown streets, both North and Southside. The torrent burst through the doors and shattered the windows of the shops on Church, Water, Madison, Lake, and every other city street within half a mile of the river. Mannequins from clothing boutiques caught against lamp posts, thought by some rescuers to be flood victims. On Water Street, only Tommy Hilfiger managed to move his stock above the flood. The owners and managers of the other businesses didn't believe the flood could reach them, or didn't have the time or muscle to get their goods out of the way.[14]

Electric power went down within minutes, and with it telephone service, radio and TV broadcasts. As at Corning, there was no communication between North and Southside, and wild rumors flew: the National Guard discovered 40 bodies in a barn (false); the Walnut Street bridge collapsed (true); the Star-Gazette

Aptly named Water Street in downtown Elmira flooded 12 feet deep.
Almost all of the businesses were destroyed, and many never re-opened
(Maxq32, Creative Commons)

building washed away (false); looters were ransacking the stores downtown (there was some looting).[15]

Short-wave radios became the jungle drums of the stricken city, separating fact from fiction, relaying reports of stranded residents, re-uniting families and friends.[16] For two days, the most reliable means of communication in Elmira was the two-way radios aboard the city's school buses. Dispatched to every evacuation center, the buses, loaded with food and supplies, parked there for the duration, their radios the only link to the outside world. Some of the drivers stayed on for forty-eight hours straight, relaying vital messages. "If it hadn't been for the radios on the buses, I don't know what might have happened," marveled school superintendent Paul Zaccarine.[17]

Not all Elmira residents left their homes before becoming engulfed by the flood. Fifteen helicopters from the New York State Police, National Guard and state Game Commission saved dozens of people from second-story windows and rooftops, carried food and supplies to the evacuation centers, and transported patients to regional hospitals.[18] Innumerable small boats manned by city police, Public Works, and citizen volunteers patrolled flooded streets and rescued stranded residents.

At its peak, the Chemung topped Elmira's dikes by more than two feet, filling both Northside and Southside to depths of up to twelve feet. Flood water raged down the river at189,000 cfs, far surpassing the 1946 record of 132,000 cfs.[19] 14,000 residents evacuated from their homes;[20] two thousand of those homes were destroyed, and another two thousand seriously damaged. Four of the five bridges that spanned the Chemung River between Southside and Northside collapsed or were so damaged they had to be demolished. Immediate economic damages came to more than $260 million,[21] In the entire Agnes event, only Wilkes-Barre suffered greater economic loss.

Despite the widespread and unprecedented destruction, despite the evacuations and desperate rescues, and despite the horrible human losses just miles away at Corning, there were no lives lost in Elmira during the flood. The Elmira task force was not caught by surprise, but in communication with Corning and the Weather Service at Binghamton, knew a catastrophe thundered toward them. DeRenzo's plan, methodical and rehearsed, even *ad libbing* as the river came over the walls, worked. Although much of Elmira was destroyed, the people survived.

The city's economic prospects after the flood were different than those of Corning. At the "Crystal City", Corning Glass Works lay at the bedrock of the city's economy, and the company took the lead in restoration of the city's vitality. Elmira, on the other hand, had no single base industry but a patchwork of already-declining factories and stores. Agnes dealt many of them a death-blow. The flood fostered urban renewal at Corning, but at Elmira permanent devastation. The city's population fell by twenty-five percent, and for four decades Elmira became more isolated, poorer, quieter.

Soon after the flood, Elmira musicians John and Marina Nickerson wrote and performed *It Sprinkled, It Rained, and It*

Poured, a song to commemorate their city's ordeal. The Nickersons donated profits from sales of the record to "Lend-A-Hand" a local organization set up to aid the victims of Agnes.[22] It begins with residents unconcerned about the impending disaster:

> *Late in the night the sirens blared*
> *But at the time no one cared.*

Then emphasizes their false sense of security behind Elmira's flood wall:

> *Because they felt proud and bold*
> *Knowing that their dikes would hold.*

THE RELAY

Wednesday, June 21, 1972
Towanda, Pennsylvania

The last river gauge on the Susquehanna River before Wilkes-Barre was at Towanda, a remote village of four thousand people. Every day LaVern Root took the measure of the river at that gauge, located in a little shack beneath the west end of Towanda's US-6 Bridge. Seventy-eight years old, Root volunteered as a cooperative observer after a long career as a power company lineman. Something to keep busy in retirement, he said. He never imagined the pivotal role he would play in Pennsylvania's greatest disaster.

Root shucked his yellow slicker before sitting at the short-wave radio in the corner of his living room and toggled the switch to transmit his data to the River Forecast Center at Harrisburg. He had just taken his 6 p.m. measure of the river. There was a quick response. "Middle Atlantic RFC, this is Patsy Quigley. Please identify your station." Weather Service intern Quigley worked late, along with hydrologists Mike Gwinner and Nick Pavick, to collect data about the floods spreading throughout the region.

"Hi Patsy. This is LaVern Root, station 01531-500 at Towanda." Root often reported to and chatted with Quigley since her internship began six months before.

"Go ahead LaVern. What are your numbers?" She added a sort of apology, "It's kind of crazy here, as you can imagine."

"I'll bet. You must have your hands full." Root knew about flooding in Virginia and Maryland, and got drenched in the rain during his last two visits to the gauge. He reported his numbers: 1.62 inches of rain since 6 a.m.; river stage 4.60 feet and rising. The Susquehanna had come up more than a foot since the morning, but was still far below Towanda's flood stage of sixteen feet.

"Got it, LaVern, talk to you tomorrow." Quigley signed off

with a warning. "We've issued a flash flood watch for the whole Susquehanna basin. Be careful up there, okay?"

* * *

By Thursday morning the river rose to eleven feet on Towanda's gauge. Root radioed his measures in to the RFC, and Quigley again answered the call. "LaVern, we're getting flooding up on the Chemung, on the North Branch too. Mr. White wants river and precip readings every three hours. Can you do that?"

He assured her that he could. "Overnight too?"

"As late as you can. Mr. White is worried about Wilkes-Barre. Your report is important. Is there anyone who can help?"

Root said his daughter Marsha knew how to read the gauge, then continued. "If it gets bad I'll call Ruth Prusack." Prusack lived on the other side of the river, and her father had been the volunteer observer before Root.

The river was at flood stage, sixteen feet, when Root took a reading at noon. At 3 p.m. it was above twenty feet, approaching the record twenty-five feet set in the 1936 St. Patrick's Day flood. But when Root tried to transmit this alarming data to the RFC all he heard was static. He had no way of knowing the radio tower at Harrisburg had been knocked out of service. There was an alternate protocol to follow if he couldn't reach the RFC, but in his seven years as volunteer observer he never had to use it.

The phone rang as Root rummaged through his desk for his Cooperative Observers Handbook. It was Gene DiLauro, Meteorologist-in-Charge at the Weather Service Office in Binghamton. Yes, that was it – Root was to report to Binghamton if couldn't reach the RFC. He thought it odd the chief would call him personally. DiLauro seemed agitated. "The radio tower at Harrisburg has shorted out, and the River Forecast Center can't receive or transmit. Until it's fixed, report to us here in Binghamton."

"Yes sir. I just tried to send my 3 o'clock. Mr. White wants me to report every three hours. Should I send it to you now?"

"Yes, send it by radio to our technician as soon as we hang up."

DiLauro confirmed the station numbers and frequency. "We'll relay it to Harrisburg over teletype with our other stations."

"Will do." Unsettled, Root paused before continuing. "I'm way over flood stage here. How are things looking? Do you know what's coming our way?"

DiLauro answered cautiously. "The computer at Harrisburg has the Chemung above flood stage all day. Corning and Elmira have dikes, so they should be okay. You'll probably see flash flooding around Towanda, and the farm areas could be in for a rough night. As for Wilkes-Barre ..." DiLauro paused, looking over the lastest teletype from Harrisburg. "Here it is. They predict a crest of twenty-seven feet early tomorrow. Frankly though, it's dicey. The rain coming tonight can't be figured in the forecasts. And some observers haven't been reporting. Probably can't get to their gauges."

Root was about to sign off when the DiLauro continued. "Mr. Root, I called you personally to be sure you understand how important your reports are. Your gauge is the last one before Wilkes-Barre. It's essential to forecasting this flood. Not only at Wilkes-Barre, but on down to Bloomsburg, Sunbury, even Harrisburg. The Forecast Center, and probably a million people down the valley, are counting on *you*."

* * *

Root turned on the CBS News after sending his 6 p.m. data to Binghamton. Anchorman Walter Cronkite opened the broadcast with a stark report:[1]

> *Good evening. We've got a major disaster along the east coast because of Tropical Storm Agnes, which has turned out to be more dangerous than anyone expected. There is other news, but none of it as important as this storm. The Weather Service believes the flooding, from the Gulf Coast to New York, is the most geographically extensive in the country's history. ... This afternoon, President Nixon declared Florida, Virginia, Maryland, Pennsylvania and New York disaster areas.*

As Cronkite named the rivers at or near all-time record floods – James, Potomac, Allegheny, Monongahela, Schuylkill, and his own Susquehanna – the importance of Root's measure of the river hit home. He never expected to be involved in a national catastrophe, but would do his damn best to get his job done.

* * *

Gusty winds kept knocking electric power off at Root's home all afternoon. It was back on now, but he had no way to know how long. Root called out to his daughter "Time for the 9 o'clock, you ready?" Marsha insisted on going with her father to check the gauge.

Root and his daughter drove through the pelting rain along US-6 until stopped by a row of flares and the flashing lights of a police car. Several drivers ahead were making U-turns. A state trooper waved a flashlight. "This road is closed, bridge flooded out. You'll have to turn around."

The beams from Roots headlights lit the road ahead. Little Laning Creek overflowed its banks and spread across the highway in a rush fifty yards wide. There was no way they could continue to the Susquehanna River bridge and the gauge. Root grimaced. "I've got to try Ruth."

The fast buzz-buzz of "line out of service" greeted Root's first try at phoning Ruth Prusack. If he couldn't get Prusack to read the gauge, how could he report the river stage to the RFC? He tried again, surprised that this time he heard the familiar ring sound. Prusack picked up right away. "Ruth, this is LaVern. I know it's late. But I need your help ..."

Together Root and Prusack worked out a strategy to read the river. By now the gauge house was submerged in the flood. As long as she felt it was safe, Prusack went onto the bridge to read the alternate wire-weight gauge. She called or delivered her measures to the local radio station, WTTC, to broadcast over the air. Root listened to WTTC at home, then transmitted the report via his short-wave radio to the Weather Service office at Binghamton. In turn, Binghamton compiled Towanda with other up-river stations and teletyped it to the RFC at

Harrisburg.[2] Five steps, the links between each of them precarious: Prusack, to WTTC, to Root, to Binghamton, to the RFC. Somehow it worked, and every three hours for the next three days LaVern Root and Ruth Prusack, with members of their families filling in, faithfully reported the crucial Towanda stage to the River Forecast Center. As the Meteorologist-in-Charge at Binghamton said, a million people were counting on them.

BOLD FORECAST

Wednesday, June 21, 1972
Wilkes-Barre, Pennsylvania

Events leading to disaster at Wilkes-Barre began to unfold early in the week. Heavy rain along the stalled mid-western cold front began pummeling upstate New York and Pennsylvania on Tuesday, a deluge fed with added moisture from the remnant Hurricane Agnes spinning up the Atlantic coast. By Wednesday some of the little creeks around Wilkes-Barre swelled over their banks, and police barricaded a few roads and bridges. Wilkes-Barre's Civil Defense Chairman Frank Townend and his executive director Nick Souchik took little notice; insignificant flooding like this was a matter for the local police, fire and public works authorities.

* * *

On Wednesday evening, the loudspeaker at the Treadway Inn's restaurant blared over the chatter of diners and clatter of dishes: *Attention please, there is a telephone call for Mr. Nick Souchik. I repeat, a call for Mr. Souchik.*[1]

Souchik and his wife were sharing a rare dinner out. "Duty calls, Olga." He stood and went to take the call.

When he returned to the table Souchik did not sit down. "It was the General. There's a new flood report in from Harrisburg. We're not sure what it means, and I've got to go in. I'll be home later." He hustled off to his office at the courthouse to huddle with General Townend and other Civil Defense staff.

A half hour earlier, as the Souchiks enjoyed their dinner at the Treadway, a new advisory had come into Luzerne County Civil Defense: the flash flood watch issued by the River Forecast Center earlier that afternoon had been elevated to a flash flood *warning*, now for the *entire* Susquehanna basin. Never before had such a warning been issued for all 27,500 square miles of the watershed. This got the attention of General Townend and his staff. But the last sentence of the RFC's advisory took away some of the sting: *Main channels of the Susquehanna, Juniata, and West Branch are well below flood levels, and no main-stream flooding is expected at this time.*[2]

A special notice was included with the new warning: O.D. White asked all volunteer observers to measure and report the river level every three hours. This, too, was a first-ever request; during floods, the RFC always before asked for readings at six-hour intervals. Had Souchik and Townend read between the lines, they might have sensed the anxiety building at the RFC. Something extraordinary was developing.

Souchik put on his raincoat, went out to read his river gauge, then came back in to report to Townend. "We're still at 4.3 feet, normal. It'll go up when we get what's coming from upriver, but I can't imagine going above flood."

The team reached a consensus. Civil Defense would stand on guard for small-scale flash flooding and monitor new advisories that came in from Harrisburg. One of the staff would stay overnight, read the gauge every three hours, and call the measure in to the RFC. If the situation took a turn for the worse he'd call Townend and Souchik at their homes.

"Looks like we have things under control," Townend walked up the basement steps beside Souchik as they left the office. "I'm going to make that trip to Ocean City tomorrow." Townend planned to attend a memorial event for the Army's 28th Infantry Division[3] in New Jersey, and was slated as a featured speaker.[4] He said he would be there "come hell or high water". This year, he meant it literally. "I'll call in every hour or so. Good night Nick, and good luck."

* * *

Thursday, June 22, 1972
Wilkes-Barre, Pennsylvania

Souchik arrived at the courthouse Thursday morning and checked his river gauge. The rain had picked up overnight, and the courthouse grounds squished beneath his feet as he walked over the grass to the gauge house. The river stood at 8.3 feet, four feet higher than the night before.

Back in his Civil Defense office, Souchik called his numbers in to the RFC, then shuffled through the stack of notices that came in overnight. The RFC's midnight forecast predicted the river would hit 5.5 feet at 7 a.m. He just measured it at over eight feet – three feet *higher* than the forecast. The forecasts were based on rain that had already fallen; they did not account for rain yet to come, so sometimes lagged behind the actual stage of the river. Still, the river stood way below flood stage of twenty-two feet.

A 4 a.m. bulletin from Pennsylvania Civil Defense headquarters in the overnight stack reported severe flooding at Harrisburg, York and Lancaster. Souchik frowned and read the report again. How could Harrisburg, a hundred miles downriver, flood *before* Wilkes-Barre? Floods went *down* the river, not up. Not only did the forecast seem wrong, the flood appeared to be going backwards. The bulletin did not explain that flooding at Harrisburg was from the extreme rains – up to eighteen inches – in the immediate vicinity, leading to local flash floods and a quick rise in the big river.

The teletype clattered again at 8 a.m. with an updated advisory, even more urgent. The RFC predicted the Susquehanna at Wilkes-Barre would hit seventeen feet by 2 p.m. and twenty-four feet after that. This was the first forecast of the river rising above the 22-foot flood stage. The bulletin concluded: *Dangerous flooding and torrential rains continue with Tropical Storm Agnes. The ground is completely saturated, and additional rainfall will cause rapid rise in many streams in the Northeast.*[5]

To Souchik and his Civil Defense staff the bulletins told a story of flash flooding of small streams, not a disaster on the Susquehanna. The predicted crest of twenty-four feet was well below Wilkes-Barre's thirty-seven-foot dikes. Public Works would lock the flood gates in at the bridge crossings, and there would be backed up sewers, wet basements, some local street flooding. It was nothing they hadn't seen before.

Another notice from Civil Defense headquarters at Harrisburg came in at 10 a.m., headlined URGENT, a *Proclamation of Extreme Emergency* issued by Governor Shapp.[6] Couched in legalese, the proclamation allowed Civil Defense to quickly hire private contractors for flood control, and authorized mobilization of the National Guard. The governor gave Townend and Souchik some tools they were going to need.

* * *

A sewer project at Plymouth, across the river from Wilkes-Barre, left a forty-foot gap in the city's riverside dikes. Plymouth Mayor Ed Burns didn't know the particulars of the forecast, but he did know that if the river got much higher it would pour through the gap and inundate the residential "flats" area of his town. Burns marshaled his public works staff, volunteers, and the National Guard to fill the breach. Over the next twelve hours, three hundred workers filled 19,000 sandbags and pushed seven thousand tons of soil into the gap, bringing the wall up to thirty-three feet.[7] By the time they finished the river had risen above the space where the breach had been. Satisfied, Burns deemed the effort a success. "That should do it."

* * *

Thick, pelting rain soaked through Frank Townend's windbreaker in the few steps from his car to a phone booth at a gas station outside Allentown. He dialed his Civil Defense office at Wilkes-Barre. "Nick, I'm halfway to Ocean City. It's raining like hell, but so far the roads are okay. What is your status there?"

"General, it's worse than we thought. The boys at the forecast center say we'll go to twenty-four feet. That's okay here in Wilkes-

Barre, but might be a problem on the west side. It's the small streams they're worried about. The creek has come over Main Street in Shickshinny and knocked out the bridge. Burns has people in Plymouth plugging that gap in the dike, and wants the 109th to help."

Townend heard radio news about the governor's declaration of emergency, and about flooding around Harrisburg and York. "OK, Nick, I'm turning around and heading back. As for Plymouth, go ahead and authorize the 109th to help Burns with his dike. We need them to start patrolling too, on both sides. See you in a couple of hours." The 28th Division meeting in Ocean City would have to carry on without him.

* * *

A new forecast from the RFC came into the Civil Defense office at 9:30 p.m. Communications director Bob Pissott tore it from the teletype machine and handed it to Souchik, who read it aloud: *Record or near-record stages are forecast for the entire Susquehanna River Valley. Towanda is 25.8 feet and rising. Wilkes-Barre 20.6 feet and rising.* The river was higher now than Souchik reported an hour ago. He swallowed hard, "The new forecast is for thirty-four feet, tomorrow morning."[8] A hush fell over the gathered staff.

Just a year before, in April, 1971, Townend and Souchik ran a drill to simulate Civil Defense's response to a hypothetical "Hurricane Judy", with a flood reaching the top of the city's dikes. The plan called for evacuation of vulnerable neighborhoods, alerts to utilities, hospitals, and businesses, mobilization of the 109th to assure public safety, contracting for supplies and equipment, and setting up evacuation shelters. Now, late in the evening of June 22, the drill Townend and Souchik ran the year before became real.

Wilkes-Barre, with its 37-foot flood wall, would be safe against a 34-foot crest. The river reached nearly thirty-one feet back in 1964, without serious damage. But thirty-four feet would slosh over the gap in the dike at Plymouth on the west side of the river. The hasty repair ordered by Mayor Burns went only to thirty-three feet, and wouldn't be high enough.

By now Frank Townend had returned and was in command. "Call Burns and tell him it's coming over his wall. He's got to get his people off the flats." Then, with military crispness, "Without delay!" Townend had another worry. The National Guard armory was in Kingston, just inside the dikes. "The 109th has to evacuate. We've got to get everything out."[9]

If you're in Plymouth or nearby low-lying areas, get out now! Plymouth must be evacuated! Townend broadcast the alert over TV and radio at 10 p.m. Police cars crawled through the city streets, bullhorning the emergency orders. *Don't panic, but leave now. Everybody out!*[10] Five thousand people – half the population of Plymouth – heeded the alarm. Many went to stay with friends and relatives in Kingston and Wilkes-Barre, not dreaming they would have to move again the next day.

At the 109th's armory in Kingston the guardsmen labored all night to load their big howitzers onto flat-bed trucks for transport to high ground. They carried their gear and equipment out by hand and hauled to safety: light arms, ammunition, supplies and records. It took more than eight hours.

* * *

At 11 p.m., Souchik turned to Civil Defense communications director Pissott, who sat at his big radio console in the corner of the sub-basement office. "Have you been able to reach Root? I need that report from Towanda." Souchik never met LaVern Root in person, but knew him as his counterpart river observer. They often talked by phone during lesser floods to share observations.

"Phones and electric are down in most of Bradford County. All I'm getting is the 'out of service' signal," Pissot fiddled with his radio controls. "I've tried seven times."

Towanda, eighty miles away, was the nearest station upriver, and key to predicting the flood at Wilkes-Barre. Souchik and Root had a rule of thumb: the river stage at Wilkes-Barre would be six feet higher and twelve hours later than the stage at Towanda.[11] They called it the "6/12 rule". Similar to the correlation between Corning and Elmira, Root's measure of the river at Towanda gave Souchik an idea of what was coming to Wilkes-Barre. "Okay Bob, keep trying. Maybe you can get it from Binghamton, or the state police. I need Towanda!"

Two and a half hours later, Pissott jumped from his console, waving a paper in his hand. It was 1:30 a.m., now Friday. "Nick, I've got that report from Towanda, over police radio, relayed from Binghamton. Your guy Root can't get to his gauge. They got someone else."

Souchik bit down on his unlit cigar as he stared at the log. Binghamton reported Towanda at 31.18 feet. "This can't be right. You sure that's what they said?" If the 6/12 rule was accurate, thirty-one feet in Towanda would be thirty-seven feet at Wilkes-Barre by noon, at the very top, even sloshing over, the city's flood walls. "I almost went crazy," Souchik later admitted.[12]

With Pissott's radio log in hand, Souchik strode across the room to Frank Townend. "General, I've got the midnight report from Towanda. It's bad, thirty-one feet up there. When that gets here it could come over the dikes! Unbelievable!"

Townend replied in his characteristically measured voice. "Your 6/12 rule doesn't always work. You know that as well as anyone. There must be something wrong."[13] He looked up at the clock, 1:45 a.m. "When do you expect a new forecast from Harrisburg?"

"Not until nine. That's almost eight hours from now." Souchik wasn't a hydrologist. He simply reported his data to the Forecast Center, and relied on their calculations to forecast the river. He knew there was complicated math and a computer involved in making a true prediction, and that his 6/12 rule was just a rough estimate. The rule didn't account for water coming in from tributaries, the "push" of water from further upriver, nor rain that fell since the last forecast. But right now the 6/12 rule was all he had, and the report from Towanda put Wilkes-Barre on the brink of disaster.

* * *

Friday, June 23, 1972
Harrisburg, Pennsylvania

Mike Gwinner, alone on overnight duty at the RFC, yanked Binghamton's report from the teletype, four pages of data compiled from the upper Susquehanna. It was 1 a.m. Most of the reports from the Chemung River were missing, but at least the data from Towanda was there. Observer Root usually radioed his data directly to the RFC, but with the radio out of service his report came from Binghamton with the batch of other stations upstate. Gwinner had no inkling of the extraordinary effort Root and Prusack made to collect and transmit their data. All he saw was the stark numbers on the print-out:

Station 01531-500 TOWANDA
Time of observation: 06.23.1972 0030 hrs.
Stage: 31.18 feet, rising.
Precipitation: 2.71 inches 12 hrs.

Gwinner called it a "miracle midnight report" when he learned what Root had gone through to get those crucial numbers.[14]

The RFC hydrologists knew the record flood at every station. But feeling that his eyes, or his memory, deceived him, Gwinner had to double check. He flipped through the Geological Survey volume of tables to the historical records for the Susquehanna. Here it was. *Towanda, 1865 to present. Record stage: 25.08 feet, March 20, 1936.* The river had just surpassed the old Towanda record by more than six feet, and was still going up! Gwinner knew the flood coming down the river from Towanda would slam into Wilkes-Barre twelve to fifteen hours later. He looked at the clock on the wall. It would hit early in the afternoon.

Gwinner was on duty overnight collecting data to run the computer forecast models when the whole staff arrived in the morning. But with the new, scarcely believable, report from Towanda time was critical. He *had* to run a new forecast immediately.

Gwinner sat at the desk of the IBM 1130 with the latest reports. He fed a stack of blank cards into the punch machine and prepared to enter the data, then stopped. This would take too long. There was usually an intern or technician to punch it all in. He couldn't do it alone while keeping up with new reports. The phones were ringing, and he already missed several calls.

Thinking quickly, Gwinner realized he could manually update the previous forecast with some of the new data. In addition to Towanda, the teletype from Binghamton reported stage and precipitation on the North Branch of the Susquehanna, Lackawanna River, and Tunkannock Creek. He meticulously penciled this new data into the algorithm, made some calculations with his slide rule, and arrived at an estimated discharge at Wilkes-Barre: 350,000 cubic feet per second.

Next, he had to translate his discharge calculation into a stage forecast – how high would the river go? The discharge-to-stage graph, the rating curve, went only as high as the previous record: 232,000 cfs in 1936. Gwinner's estimated discharge was not only off the chart, but off the paper the chart was printed on. Using his architect's curve he sketched an extension of the graph, and read down. The river at Wilkes-Barre would rise to forty-two feet: nine feet higher than the "insurmountable" 1936 record, five feet higher than Souchik's "6/12 rule", and five feet over the top of Wilkes-Barre's flood walls.

A number this extraordinary had to be verified. When there are no major tributaries between two stations, the stage at any given station can be correlated to the stage at the nearest station upstream. This is called the *stage-stage*, or *crest-crest*, relation, a more sophisticated version of Souchik's "6/12 rule". A stage-stage chart is based on decades of observations, and can serve as a "sanity check" against predictions made by hydrologic modeling. If ever a sanity check was needed, it was now.

Gwinner pulled the Towanda-to-Wilkes Barre chart from the RFC's files. As with the rating curve, the stage-stage chart went only as high as the previous record. Again he extended the graph: thirty-one feet at Towanda would be forty-one feet at Wilkes-Barre. It wasn't "by the book", but good enough to verify the forecast. Gwinner stared at the number he had just written. "God help them."

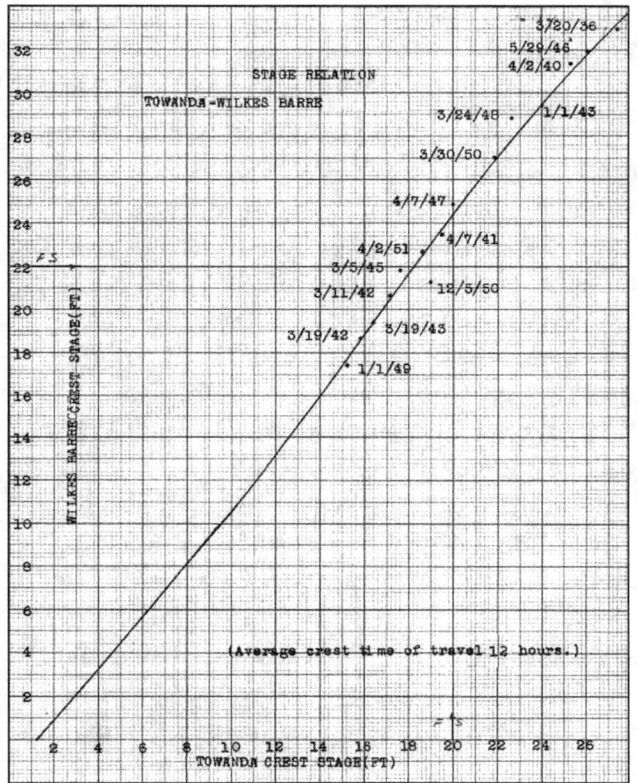

A 1960s "stage relation" chart for Towanda-to-Wilkes-Barre, showing the 1936 flood as the maximum. Gwinner's forecasted crest for Agnes was off the paper the chart was printed on (NWS, River Forecast Center)

Standard protocols required that the RFC teletype its forecasts to the regional Weather Service offices, in this case Wilkes-Barre/Scranton. They, in turn would distribute the forecast to local authorities and media. Civil Defense headquarters would also get the forecast when flooding was predicted, then pass it on to their county offices. But this was not an official forecast, and not a standard situation. There was no time to follow protocols. Wilkes-Barre had to be warned.

Gwinner flipped open the roster of volunteer observers and found the phone listing for Wilkes-Barre. He pressed his Line 1 button and dialed the number.

* * *

Friday, June 23, 1972
Wilkes-Barre, Pennsylvania

It was 2 a.m. Cigarette haze hung heavily in Civil Defense's windowless sub-basement headquarters. The phones rang urgently, teletype chattered, short-wave radio squawked. Staff huddled in small groups to pore over maps, engineering diagrams, and notebooks full of disaster response plans. Their haggard faces and rumpled clothes revealed they hadn't slept since the day before.

Souchik took Gwinner's call. "Mike! I was about to call you. We're getting nervous up here. I just came in from my gauge, 28.9 feet and rising. The report from Binghamton has Towanda at thirty-one feet. I figure we'll be six feet higher in twelve hours, so thirty-seven feet. That's right at the top of our walls. Tell me I'm wrong!"

"Listen, Nick. You're not wrong. I've just run the numbers. This is back-of-the-envelope, but the best I can do." Gwinner blurted it out. "I put Wilkes-Barre at forty-two feet, by Saturday morning."

Souchik was silent for a moment. Then his response exploded into the phone. "Goddamn! That's impossible. You know our dikes are thirty-seven feet, right? Are you saying we'll go *five feet* over the top? Are you sure?"

"Yes Nick. I'm sure. I'm forecasting forty-two feet, but can guarantee forty-one. It will go over your walls between ten and noon this morning. Believe me, Nick. *You've got to get everyone away from the river.*"

"Mike, General Townend has to hear this himself." Souchik handed the phone to his boss.

* * *

Gwinner's bold forecast[15] that night was hailed as saving hundreds, perhaps thousands, of lives.

Direct personal contact was made by the RFC to the Civil Defense unit at Wilkes-Barre, ensuring that the seriousness of the situation was fully understood. Special mention is made of this forecast

because it was the instrument responsible for triggering the mass
evacuation in the Wyoming Valley, averting the possibility of an
unthinkable loss of life.[16]

"What," mused the Corps of Engineers, "if the evacuation of
Wilkes-Barre had been delayed?"[17]

"Unquestionably," answered NOAA, there would have been a
"disaster of unimaginable magnitude."[18]

EVACUATE!

Friday, June 23, 1972
Wilkes-Barre, Pennsylvania

General Townend set the phone back in its cradle and turned to face his staff. A tense, expectant calm spread over the room. "That does it. Harrisburg says the river will be over the wall by noon, maybe six feet over by Saturday. It's hard to believe, but he seems sure."

Everyone began talking at once, the calm broken. Townend raised his hands. "People, people! Listen up!" There was no doubt he was in charge, the voice of authority. "Here is our situation. We've got to assume this forecast is correct. If it is, all of downtown Wilkes-Barre will go under. Kingston, Forty Fort, Swoyersville, Edwardsville – anywhere in the county low to the river. We have to evacuate the whole area."

Still, Townend clung to hope that Gwinner's forecast was wrong. Everyone *knew* the walls were high enough, and it was hard for some of them to let go of that long-held belief. If Souchik's "6/12 rule" was right they might have a chance. Townend turned to Luzerne County Engineer Bernie Gallagher. "What's our sandbag situation? If we can raise the dikes by a foot or two, can we hold it back?"

Gallagher looked at the notebook in his hands. "I've got twenty thousand bags in the city. My crews are stacking them around the water plant now. And we've been hauling sand to the bridges since the afternoon." Gallagher paused. "General, that won't be enough, and I haven't got the manpower."

"How many bags will we need?"

"We ran this in the drill last year. But now we're talking about miles of dikes, not just the low spots. It's going to take a million bags. More or less."

Souchik interrupted. "I've already called into Eastern for more." For several years Pennsylvania Civil Defense warehoused sandbags at regional depots for distribution when needed. Eastern Region had a stockpile near Harrisburg. "But everything has already been sent out to other areas.[1] They might be able to scrounge some from Virginia."

Townend rubbed his eyes. "A million bags. Jesus. Okay, keep trying, get what you can. For manpower, we'll put out a call for volunteers. It's going to take ten thousand men." As he had at the Battle of the Bulge twenty-seven years before, General Frank Townend was about to command an army.

The battle plan called for a two-pronged defense. They would assume the river would come over the dikes, and evacuate their city. At the same time, hoping for the best – that Gwinner's dire forecast would not come true – they would buttress the flood walls.

At 2:30 a.m., less than an hour after hearing Gwinner's forecast, Townend broadcast a special bulletin over WILK radio:

> *The flood situation is worse than we thought. ... By Friday afternoon the river will be at thirty-eight, maybe forty feet. We are going to try to raise the dikes with sandbags, if we can get them. ... This is not a job for four hundred, six hundred, or even a thousand men, but will take at least ten thousand people. Be at the dikes at 6 a.m. It's the only way it can possibly work.[2]*

Townend's statement was recorded and broadcast on radio and television over and over in the coming hours.

* * *

Gwinner predicted the river would reach the top of the walls by noon. That gave Civil Defense ten hours to prepare. Souchik leaned close to Townend. "That's eighty or a hundred thousand people. General, we have to get them moving."

Townend, accustomed to command in battleground conditions, was more cautious. "No, Nick. It would be chaos. People would panic in the dark. We'll stick to the plan, like we drilled last year. Get the hospitals cleared first and give the Red Cross time to set up. At daylight we'll evacuate the rest."

There was more than enough to do, an impossible amount of work in a few short hours. One of Civil Defense's strongest assets was its network of personal connections. Wyoming Valley families had lived there for generations. Politicians, administrators, media and business owners knew each other well. When Civil Defense staff directed action or called for services, they weren't talking to strangers, but to people already known and trusted: brothers at the lodge, members of their church, sisters-in-law or cousins. Personal relations proved to be a tremendous advantage in the hours and days to come.

Evacuation of unprotected low areas on the west side of the river – Plymouth, Kingston, West Pittston, Swoyersville – began earlier in the evening. Now, in the face of the dire new forecast, the evacuation zone widened. Residents carried what they could to the second floors of their homes and packed a few belongings into their cars. They figured they would be back the next day, never thinking that the second floors of their homes and businesses wouldn't be high enough. A week would go by before they returned to see what could be salvaged.

The Red Cross called out its legion of volunteers to prepare eighty-one shelters at schools and other public buildings on high ground around valley. Trucks laden with cots, blankets, first aid supplies and packaged food trundled from warehouses through the dark, rainy streets. Many churches and synagogues likewise opened their doors, quietly calling in parishioners to help house and feed the expected evacuees.

Engineers and linemen were awakened to shut off electric and gas service in the low-lying areas of Wilkes-Barre and the surrounding river towns. Much more complicated than throwing a master switch, the technicians swarmed to switch centers, transformers and pump stations throughout the valley. It would take hours.

Townend personally called Mercy Hospital. Located only blocks from the river, Mercy was one of the city's main health care providers. It would take time to move the patients to safety on high ground, and crucial to get their evacuation started.

Dr. William Pearlman at Mercy took Townend's call. Pearlman was on duty as attending physician in the emergency room, and organized the evacuation. He told Townend that Wilkes-Barre General, Wyoming Valley Hospital, the big Veterans Hospital, and several smaller facilities were already prepared to receive his patients. Now, hearing the forecast, he would move them immediately.

* * *

At 3 a.m. Nick Souchik went out again to read the river, then sloshed back to the Civil Defense office. "It's come up three feet since midnight! 32.3 feet, almost to the 1936 crest. At this rate we'll be over the wall way before noon!" Hope that Gwinner's forecast was wrong faded. The river was rising hours ahead of the forecast.

Townend resigned to the inevitable. They couldn't wait until dawn. At 4:20 a.m. he went back on the air, speaking calmly:

> The best information we have indicates that the river will come over the dikes sooner than expected. ... The people in Wilkes-Barre and Kingston must get to high ground. If you are hearing this broadcast you still have time.[3]

Townend went on to specify the neighborhoods to be evacuated, and places residents could go. He would blow the city's sirens if – and when – the river came over the dikes, when all residents must drop whatever they are doing and *go*.

The sandbagging effort would go on:

> We are going to try to beat the river, to keep it within the dikes. We again appeal for people to report at 6 o'clock to raise the level of the dikes. We're not going to give up![4]

Townend and his Civil Defense staff still grasped to a slim hope the forecast was wrong. Maybe the rain would stop, or maybe there was an error in the calculations. If the flood eased a little, then maybe, just maybe, with a little added support, the levees and walls would hold. County Engineer Gallagher told Townend that they couldn't possibly build a sandbag wall high enough to keep a 41-foot crest from inundating the city, but if the crest were only thirty-seven feet, or even thirty-eight, it just might work. At least they could buy some precious time.

* * *

Friday, June 23, 1972
Wilkes-Barre, Pennsylvania

Bill Shock's Aunt Katherine woke him from a deep sleep. He had a phone call. Shock tried to focus on the little alarm clock at his bedside table. 5:30-a.m. "Geez, it's too early. Who is it?"

"Someone named Ellen Rogers."

Suddenly wide awake, Shock remembered Rogers from the party on Public Square the previous Saturday. Why would she be calling him, and at this hour? He tugged on a pair of shorts and stumbled to the kitchen. His aunt seemed worried. "We were up anyway, listening to the news."

Shock picked up the phone. Rogers's words tumbled out. "Haven't you heard?" Only now did he realize his aunt's troubled expression. Something was happening. Rogers continued. "There's going to be a flood. This guy on the radio, a general or something, says it will be the worst ever. The whole city could be flooded!"

He still didn't understand why Rogers called him. Last night Shock's aunt and uncle talked about flooding in Plymouth across the river, but nothing about Wilkes-Barre.

Rogers explained. "They're going to try to make the wall higher by piling up sand, and want volunteers to come down and help." She paused to take a breath. "I thought you might want to go. They said to bring shovels."

* * *

At Nesbitt Hospital in Kingston, Alice Horowicz, seventy years old, was recovering from surgery. She opened her groggy eyes when the nurse came to her room.

"Okay, Alice, we are going to go on a little trip." The nurse tried to be cheery. "You're going to pay a visit to Misericordia College." The small Catholic school up the hill in Dallas had set up a dormitory for convalescing patients.

"What? Why?" Horowicz didn't know about the expected flood. She hurt all over, and dreaded the move.

The nurse bustled around the room, collecting Horowicz's things. "Orders. They say we might get flooded here. Come on now Alice, let's get you onto the gurney."

Draped in a white sheet and her eyes closed, Horowicz grimaced as the nurse wheeled her through the hallway. They passed through the emergency room door, and Horowicz felt the legs of the gurney fold beneath her as it slid onto the gimbals of the waiting vehicle. As she opened her eyes she saw she was not being loaded into ambulance, but into a hearse. "No! Nurse! Where are they taking me? I'm not dead yet!"

There weren't enough ambulances to carry the hundreds of patients from hospitals and nursing homes to safety up the hill. John Cuff, president of Luzerne County's association of funeral homes, realized that the gimbals for sliding a casket into a hearse were the same as those for loading a gurney into an ambulance. He called Civil Defense and volunteered the services of his members to transport hospital patients.[5] The passengers may have been shocked, but all safely arrived at their temporary new quarters.

* * *

At Wilkes-Barre General Hospital, barely above the flood on Franklin Street, Dr. Lester Saidman surveyed the ambulances and funeral cars queuing up at his emergency room doors. As chief of the ER he arrived at 6 a.m., alerted to the influx of patients from the other hospitals. There would more emergencies than on a usual day: heart attacks from stress about the flood, volunteers injured in the sandbagging operations, even near-drownings. By 8 a.m. the ER was filled to capacity.

Dr. Saidman brought his 15-year-old son Bruce to the ER with him, thinking the teen could serve as a volunteer orderly. "Here, put on this white coat, and help get the folks from the nursing home inside. We've got to speed this up." It was the first time young Saidman ever donned any kind of medical clothing.

An old city bus filled with nursing home patients pulled up. Bruce helped them into wheelchairs and tried to find a place for them inside. Many patients seemed dazed. Cots, gurneys, and wheelchairs filled the emergency room and spread into the lobby, the lounges, and the hallways. By noon the number of patients at Wilkes-Barre General more than doubled, from 310 to 800.

Dr. Saidman, a veteran Army doctor, went into triage mode. "A lot of the doctors had been in World War II, so we did what we had done in the military, making snap decisions based on the severity of the patients' condition – whether to hold them, discharge them, or send them in for immediate surgery. We had to think on our feet all the time, and then we sweated it out."[6]

Electric power and phones were down throughout the valley, but emergency generators at Wilkes-Barre General kept the lights on and equipment running. The radio used for incoming ambulances was the only means of communications. Young Saidman wheeled another patient into the lobby when a young woman sitting at the radio called to him. "Bruce, I need to take a break, can you handle this radio for a few minutes?" It was Nan Schelowitz, daughter of one of the other doctors and a few years older than Saidman.

"I guess so. What do I do?"

Schelowitz gave Saidman a two-minute crash course on radio operations. "This dial changes the frequency, for the different people we need to call. Here." She handed him a list identifying the call signs of city and private ambulances, other hospitals, state and local police, fire departments, Civil Defense, the National Guard, and a dozen others. "Mostly though, they are calling us, on our frequency. Right now I am waiting to hear back from the Army depot at Tobyhanna about some supplies. Be sure to write everything in the log, I mean everything. We can take turns bringing messages around the hospital."

For the next twenty hours, teenagers Saidman and Schelowitz were the *de facto* communications center amid the chaos at Wilkes-Barre General Hospital. Most of the radio traffic came from ambulances and helicopters bringing patients, and shuffling supplies and medicines among the city's medical centers. They also heard the chatter from police, fire, and Civil Defense about the rising flood and evacuation of the valley, of the river breaking over the wall, of stranded victims and desperate rescues. "It was a tremendous responsibility for a 15-year-old, and very exciting," Saidman later told a reporter.[7]

As Dr. Saidman and his son toiled all of Friday and into Saturday at Wilkes-Barre General, the Susquehanna ravaged their riverside home in Kingston, leaving only a heap of soggy rubble. Many of the valley's doctors lived in the upscale homes along the river and in the well-to-do neighborhoods on the west side, areas where the flood punched its hardest. Dr. William Pearlman of Nesbitt Hospital, who lived on River Street in Wilkes-Barre, spoke for many of his colleagues. "My home and office are in a landfill. My books and journals have been lost. But I am completely confident in the Valley's ability to get back on its feet."[8]

* * *

Spurred by Frank Townend's middle-of-the-night broadcasts, tens of thousands of Wyoming Valley residents threw clothes in a suitcase, gathered their important papers and medicines, locked their doors, and headed to high ground. As volunteer sandbaggers streamed down toward the river, other residents made their way up the hill. By dawn tens of thousands of them were safe above the rising Susquehanna. The panic Townend feared did not materialize.

Still, not everyone heard the early warnings. And not everyone believed the forecast. Carl Schwab, a 32-year-old newspaper employee, lived with his family of six in a duplex home on Academy Street. On Thursday evening he attended a banquet honoring his paper's delivery boys. "A lot of people were calling the hall and asking that their sons be sent home. They were getting nervous about the flood. But most of us weren't scared – we'd seen the water come up before."[9]

Schwab watched Friday night's 11 p.m. news on WILK, when the Susquehanna River was forecast to hit 34 feet, but was fast asleep when Townend broadcast his warning to evacuate. Early Friday morning Schwab was surprised to get a phone call from his sister Gladys. She lived up the hill in Mountain Top. "Carl, you've got to get out of there! Come up to our house until this is over, okay?"

Schwab's home was a thousand feet from the river. He couldn't believe they were in danger. "Thanks Gladys. I think we'll be okay here. We've never flooded before. But if they blow the sirens like the guy said, I'll bring Dee and the kids up for the day."

"Alright. But we're worried about you. Remember what mom and dad told us about the flood in 1936. The news on TV says this might be even worse."

By 8 a.m., dozens of people bustled along the sidewalk in front of Schwab's home. Many toted suitcases or duffel bags. Shouting parents tugged their balking children, imploring them to hurry. Urgent bulletins on TV and radio reported flooding across the river in Plymouth and Kingston, and warned residents of Wilkes-Barre to wait no longer.

The urgency of the fleeing residents of Wilkes-Barre was infectious, and Schwab at last realized it was time to get his family away from the river threatening their city. He rushed his wife and children into their station wagon and set off to his sister's home in Mountain Top. They packed no clothes and moved no furniture. Their pet parakeet remained in its cage in the living room. The family didn't imagine the rising waters of the Susquehanna could ever reach it.

Traffic crawled on the Schwabs' route up the hill, and on all the other arteries leading out of town. "It was just a string of cars, as far as you could see." A hundred thousand residents were on the move. About a third of them, like Carl Schwab and his family, stayed with family or friends. Some lodged with Good Samaritans who opened their homes to strangers. Many more found their way to one of the shelters set up by the Red Cross, the Salvation Army, and the area's churches. Almost everyone figured they would return home soon. "We had only our clothes, and thought we would be at my sister's for a day. Well, that didn't happen."[10]

SANDBAGS

Friday, June 23, 1972
Wilkes-Barre, Pennsylvania

General Townend called for ten thousand *men*, not imagining the motley recruits that assembled at the river. Yes, there were burly construction workers, grizzled miners, and paid contractors. But what Townend did not expect was families – fathers, mothers and their children, ready to pitch in, shovels in hand. Platoons of volunteers from churches, the Lions, Elks, Rotary, Masons, Kiwanis, and Knights of Columbus joined the makeshift army. A troop of forty carnival workers, their performance rained out, dug in at Pickering street.[1]

Most of all, the youth of Wilkes-Barre's came to the flood wall to defend their city. High school and college kids, boys and girls, the hippies and hooligans rousted from Public Square and derided in the Times-Leader just a week before answered Townend's call. In their cut-off jeans and sneakers they streamed from their homes – many of which would be destroyed – to Wilkes-Barre's dikes to fight the inexorable Susquehanna. "They were unbelievable, tremendous. Young people working on the sandbagging actually used their bodies to shore up the pile while more bags were put into place,"[2] recalled city manager Leo Corbett.

Impossible to fortify the entire length of flood wall – there wasn't enough time nor enough sandbags – the defense focused on weak spots, around bridges, and places where the wall had settled or eroded. Townend knew it was a long shot; in his 6 a.m. broadcast he put the odds of holding back the river at only one in four. But if they could raise the wall a couple of feet, if the river didn't come as high as Gwinner predicted, there was at least a chance to save the city.

* * *

Will Shock and Ellen Rogers joined a thousand other volunteers at the Market Street Bridge, one of the points of assembly announced on the radio. Misty drizzle hung in the heavy air, punctuated by short bursts of showers as the trailing rain bands of Agnes scudded past. Shock was glad he wore his heavy work boots and yellow rain slicker.

A man in blue Public Works coveralls stood on the hood of a truck with a bullhorn. "QUIET PLEASE! Okay. We've got to stop the river from coming over the dikes. Try, anyway." The volunteers listened, straining to hear over the roar of the river and metallic groan of the corrugated steel flood wall. "Sand is piled up where we need you to fill sandbags. The bags are there, and more coming."

The Public Works man gestured toward the area where Shock and Rogers stood among a hundred other volunteers. "This group, go down to Ross Street." Then, pointing to another group and another direction, "And this other group, up to North Street, by the courthouse. The rest of you stay here at Market Street."

The volunteers shouldered their shovels and began walking toward their assigned areas when the Public Works man again lifted his bullhorn. "WAIT! One more thing. This is important!" The volunteers turned around. "If the water breaks over the wall, the city will blast the sirens, seven times. That's the signal to stop work and get away from the river. If you hear the sirens, stop work immediately, and get away. So be alert for the sirens. Does everyone understand?" There was a murmur of assent, and some thumbs up. "Okay then. Let's get going!"

The intersection of Ross Street and Riverside Drive and was three blocks away. When Shock and Rogers arrived they found men pushing gravel and mine culm up against the wall with bulldozers and front-end loaders. Shock recognized one of the operators. "Howie, what are you doing here?"

"Hey Bill! Come to help? I'm supposed to pile this stuff against the wall so you guys can stack the bags on it. My boss says the river will come over anyway. But we'll give it a try." Howie revved the diesel engine of his loader and scooped up another bucket of culm.

The corrugated steel flood wall at Ross Street stood five feet tall atop an earthen levee, ironically blocking the volunteers' view. Shock, Rogers, and the others couldn't see the river they came to fight. Yet their foe made its looming presence known by the earthy smell of its silt-saturated waters, and the creak of the wall as the river battered against it.

The mood was not, yet, one of desperation. One young volunteer put a Led Zeppelin tape in his boom box and blasted *Stairway to Heaven* to the sandbaggers. Rock and roll competed with the revving engines and reverse beeps of the loaders. The soundtrack somehow seemed apt. This wasn't a party on Public Square, but the volunteers were among friends, and pulsed with youthful energy. They gathered to fight the flood together.

A half dozen pallets of burlap sandbags stood in a line at the Ross Street intersection, and two big heaps of sand were piled nearer the wall. Rogers stood by one of the sand piles, her shovel in hand. "Okay, how are we supposed to do this?" No one was in charge.

A grey-haired man in a cap with a Caterpillar logo rushed over. "Hold the bag open, girl. And you." He pointed to Shock. "Start filling it. One of the men will take it to put on the wall."

"You think I can't use a shovel?" Rogers narrowed her eyes. "Will, hold the bag open for me."

"Okay, okay." The Caterpillar man backed away, muttering to himself. "This ain't gonna work."

After watching the others, Rogers filled each bag to the top, leaving only a fringe of burlap as a grip. No one told her the bags should be only two-thirds filled, so the flap could be folded over and sealed.

Volunteers slipped in the muddy ground as they formed a "bucket-brigade" to pass the over-filled bags to the wall. There wasn't much chatter, only the grunts of the stackers as they hefted the fifty-pound bags higher and higher. No one told them to build the stack as a pyramid against the wall, rather than a vertical pile.

All of the bags on hand were filled and stacked within the first hour, yet didn't even reach the top of the wall. No longer smiling, their faces and clothes smeared with mud, the volunteers had no instruction about what to do next. Shock rubbed his sore muscles and stretched. The Led Zeppelin tape ended.

A panel truck trundled down Ross Street and skidded on the wet pavement as the driver stopped at the intersection. Lettering stenciled on its side read *Hotel Sterling*.

The driver stepped out. "Who's in charge here?"

One of the older men said no one from the city was there, but he ran the heavy equipment.

"That's good enough for me. I've got a couple thousand pillow cases. Can you use them?" A dozen volunteers flocked to the truck, pulled out the pillowcases, and stacked them by the sand pile to be filled. They would do nicely.

There were not enough sandbags anywhere along the wall. When Townend heard about the shortage he went back on the air: *Paper bags, plastic bags, pillowcases, anything that will hold sand ... put these items outside your door and turn your porch light on!*[3] Volunteers ran up and down streets throughout the city to retrieve bags for sand: Christmas shopping bags from city merchants, mail sacks from the Post Office,[4] even old shirts with their sleeves tied closed. Pocono Downs race course sent thousands of feed bags, emptied of oats and alfalfa.[5] Garment factories and tailors churned out sandbags instead of curtains and clothing. Soon after 10 a.m. a National Guard flight with 100,000 bags scrounged from the Virginia warehouse landed at the airport, where city trucks stood by to bring them to the wall.

And still, there were not enough. It was too little, too late.

* * *

Back at Civil Defense headquarters, Souchik noticed it first. By 8:00 a.m. the accumulated rain, though diminished to a grey drizzle, pooled around the old courthouse and inexorably worked its way down the basement walls. It found minute cracks in the masonry foundation and crept its way inside to puddle around Souchik's feet. He looked up to find Townend. "General! We have a problem here!"

The water coming in was seepage from the rain, not flooding from the river. But as they stared at the growing puddle Townend and

Souchik realized that if the river topped the wall their headquarters in the sub-basement would fill to the ceiling. An office below ground might be a good place to hide from military attack, but the very worst place to be in a flood.

Townend again called for quiet in the room. Even as he spoke, the puddle expanded to the middle of the floor. "People! We have to move!"

Communications gear and other equipment in the Civil Defense office cost $300,000. Most of it would be ruined if it got wet. "Let's get this stuff upstairs!" The entire staff scrambled to disconnect the scanners, recorders, phones, typewriters and other devices to carry one flight up to an empty space in the first basement. Charts and papers were hastily jammed into boxes, and maps rolled into four-foot-long tubes. The big radio control panel – the most expensive equipment in Civil Defense's arsenal – was too big to move and left behind. Communications expert Pissott pulled the power cord and connections from the teletype machine – their crucial and secure link to Civil Defense headquarters in Harrisburg, and to the River Forecast Center – and carefully carried it up to the new temporary office. Of all their equipment he wanted this piece reconnected first.

* * *

Friday, June 23, 1972
Harrisburg, Pennsylvania

O.D. White arrived at the River Forecast Center at 6:00 Friday morning. He looked over Gwinner's penciled 2:00 a.m. forecast for the crest of the river at Wilkes-Barre. "This looks good, Mike. It took courage to make this prediction."

Quigley was already punching the overnight data to run the computer model for a new forecast. By now, twenty-three of the sixty observers in the Susquehanna basin had fled for their lives. Volunteers that could reach their instruments continued to send data every three hours, including Souchik at Wilkes-Barre and the Prusack/Root relay from Towanda.

Hydrologist Lars Feese gave the IBM 1130 instructions to forecast every station on the Susquehanna in Pennsylvania, from Towanda to Marietta. He hovered over the printer, waiting for it to finish, then brought the forecast directly to White. It was a few minutes past 7:00 a.m.

White shook his head. "Record crests all the way down. Towanda, Wilkes-Barre, Danville, Sunbury, Harrisburg." The new run put the Wilkes-Barre crest at forty-three feet, Saturday morning – a foot higher then Gwinner manually forecast only a few hours before.

In most areas the RFC's predictions were relayed to local media through regional Weather Service offices. The RFC was a "man behind the curtain", not known to the public. But here in Pennsylvania's capital the RFC gave its forecasts directly to local media, and was cited every day in the Patriot-News. White was often asked to make a statement when the river ran high. This morning, Harrisburg radio station WHP set up a 7:15 telephone interview about the continuing flood. It was a familiar routine for White. The radio producer called a few minutes before the interview to go over the questions, reminded White of the time constraints, then counted down until White would be on the air. "On my mark. Five, four, three, two ..."

At that very moment, rain water from overflowing storm sewers infiltrated Pennsylvania Electric's underground conduits and short-circuited the mains, tripping breakers at the regional substation. Power in downtown Harrisburg blinked out. Overhead lights at the RFC flickered off, the radio and teletype fell silent. Quigley stumbled from the windowless computer room at the center of the office. "The computer just died!" The IBM 1130, indispensable for forecasting the flood, useless.

Grey light filtered through the windows, casting a dim pall throughout the RFC. The hydrologists and staff stared at one another. "*Now* what do we do?"

White stood from his chair, still cradling the phone to his ear. To his surprise, he could still hear the WHP reporter at the other end. The phone line was still alive. After a moment's stumble he completed the interview. "We are forecasting record crests on the Susquehanna all the way from the New York state line down to Chesapeake Bay. Here at Harrisburg, we expect the river to hit thirty-three feet, by Saturday evening."

"But Mr. White, hasn't the river gone down since it peaked on Thursday?"

"Yes, a little, for a short while. Don't be deceived. We're not out of danger. The rise yesterday was from extremely heavy rain right around Harrisburg. The higher crest we predict for tomorrow is coming from up north, and will affect the entire length of the river. Flooding here at Harrisburg will be even more extensive than it was yesterday. My office is working with Civil Defense to identify areas that should be evacuated."

The interview ended. White turned to the disaster facing his office. The RFC was paralyzed without power for the computer, radio, and teletype. The phones lines worked, but the electronics

Volunteer sandbaggers ironically couldn't see the river they came to fight, pictured here at South Williamsport (Courtesy Bryson Leidich Photography)

controlling the switching system did not. White's hydrologists and staff could make outgoing calls, but incoming calls didn't ring even though a caller was on the line. Every few minutes they picked up each phone to see if there was anyone there.

White's first priority was to get the new forecasts to Civil Defense and the regional Weather Service offices. He turned to hydrologist Lars Feese. "Can you write this forecast up as a bulletin? Civil Defense has a back-up generator. If we hand-carry it, they can send it to their county offices and the local Weather Service."

Lars Feese, with his hand-written bulletin in his grasp, stepped out of the RFC office to the elevator and reflexively pushed the "down" button. Only when it didn't light did he realize that the elevators, of course, couldn't operate without power. He would have to walk the eleven flights down. His shoulders sagged. And eleven flights back up.

It took Feese twenty minutes to walk the three blocks to the Dauphin County Courthouse. As at Wilkes-Barre, the state Civil Defense was headquarters in the basement. Unlike Wilkes-Barre, the courthouse was well above the river and not in danger of flooding. A backup diesel generator kept the Civil Defense office lit despite the city-wide power outage. The RFC's bulletin with the ever-more-dire forecast went out over Civil Defense's teletype at 9:30 a.m.

* * *

Friday, June 23, 1972
Wilkes-Barre, Pennsylvania

By 9 a.m. it was apparent that Luzerne County Civil Defense's move from the sub-basement up one flight to the first basement wasn't enough. If the river crested over the dikes, flood waters would fill this space too. Again, the staff and volunteers lugged the documents and delicate equipment up another flight of stairs, this time to the county sheriff's office on the first floor. Again, Pissott disconnected the teletype and carried it to the new space.

The timing could not have been worse. Civil Defense headquarters dispatched the RFC's new forecast during the twenty minutes Wilkes-Barre's teletype was disconnected. The update finally rattled in a few minutes after 10 a.m., buried among other messages from Civil Defense Headquarters, State Police, and National Guard. Several minutes more went by before Pissott read it.

"General, we have an updated forecast from Harrisburg." Pissott's hand was shaking as he handed the bulletin to Townend. "Now they're saying forty-three feet."

If Gwinner had not made his impromptu 2:00 a.m. emergency forecast, this notice from the RFC would have been the first Wilkes-Barre learned of the impending over-top of its flood walls. The news would have come far too late for an orderly evacuation of the Wyoming Valley, or to marshall volunteers to buttress the wall.

Townend's heart sank. Six feet above the wall, and no hope his army of sandbaggers could stop it. His voice weary, resigned, "Bob, get White at Harrisburg on the phone. I need to hear this from the horse's mouth."

Souchik listened in on Townend's brief conversation with O.D. White. Half again as much water was coming down the Susquehanna than had ever come before, and the model, based on historical records, was extrapolating into uncharted territory. Nevertheless, forty-two feet was a sure thing. To make matters worse, White told Townend about the power outage in Harrisburg. Manual calculations without their computer would delay further forecasting.

"General," Souchik insisted, "Blow the sirens, get everyone out!"

Perhaps Townend remembered taking a stand against an overwhelming foe in the War. Perhaps he clung to common knowledge that the dikes couldn't be topped. Perhaps he was hoping for a miracle. He would stick to his plan. "Not yet, Nick. I said we'd blow them when it comes over. That's what we'll do. There's still time." He paused. "It won't come over all at once."

Pandemonium can barely describe the scene at Civil Defense's temporary headquarters in the sheriff's office. Townend, Souchik, and their staff handled emergencies not only in Wilkes-Barre, but in all the other river towns in the county. Plymouth, Kingston, and Swoyersville were already flooding, with evacuations and rescues underway. Soon, the river would rip into Edwardsville, Hanover, Nanticoke, Forty Fort, Plains, Wyoming, Plains and Pittston. Vying for Civil Defense's attention: the mayors, Public Works, local and state police, fire and ambulance, Red Cross, Salvation Army, National Guard, gas and electric utilities, private contractors, hospitals and nursing homes, newspapers, radio and television stations, major industries, small business owners, and a hundred thousand residents. It was not humanly possible for Civil Defense's six staff and a dozen volunteers, crammed into the improvised space, to keep up. Urgent calls came too fast: *Six people stranded on a roof in Plymouth. More sand at Market Street. Electric has downtown shut off. Another bulletin from Headquarters. Not enough ambulances at Nesbitt. The sandbags from Virginia just landed. Looters in Pittston. Bridge out in Edwardsville. Red Cross wants a truck at GAR High School. The 109th has everything out of the armory.*

At 10:30 a.m. Sergeant John Traylor drove the last truckload of supplies though the flooded yard of the 109th's Field Artillery armory in Kingston. The water rushed four feet deep inside the building, but all of the arms, equipment, records and supplies had been safely relocated. For the past eight hours the guardsmen worked to protect their own headquarters and equipment, and did not engage in the battle to buttress the dikes. The first platoons of guardsmen arrived at the Market Street Bridge at 10:45. They found a thousand soggy, exhausted volunteers still filling sand bags and stacking them a foot or so higher than the top of the wall.

OVER THE WALL

June 23, 1972, 11:00 a.m.
Wilkes-Barre, Pennsylvania

Ellen Rogers stood at the dwindling piles of sand and pillowcases. She was drenched from the drizzle and sweat, her sneakers sodden, her dark hair matted, her clothes plastered with sand and mud. "It's not working, Bill."

The flood wall at Ross Street stood outside a bend in the river. Centrifugal force of the water around the bend tilted the river up against the wall, and compounded the pressure of the flood. The increasing height of the water and pressure took their toll, knocking sandbags loose and splashing over. Some exhausted volunteers drifted away, while others stayed to plug the holes and stack the bags another row higher.

At 11:00 a.m. a ten-foot row of sandbags collapsed off the top of the pile. Volunteers screamed as the bags, over-filled, unsealed, and improperly stacked, spilled their contents and crumbled from the makeshift rampart. Flood water spewed through the breach as if from a giant fire-hose. "RUN!"

Dozens of panicked workers retreated from the floodwater bursting through their wall. The man with the Caterpillar hat turned to Shock and Rogers. "It's over! Go, now!" Then, as he turned away, added, "You, all of you, have been great."

Shock was confused. The Public Works man said the sirens would blow when the volunteers had to retreat. But there were no sirens, even though flood water cascaded down the levee and the volunteers retreated. "I'll stay til they give the order". His voice could barely be heard over the roar of the river and shouts of other volunteers.

Rogers handed her shovel to Shock. "Not me, I'm out of here! Good luck." She turned and splashed away through frothy brown water now ankle deep and running swiftly in the road.

Leo McCormick, assistant director of Luzerne Civil Defense, was at Ross Street to check on the sandbag operation. He knew this weak spot could be one of the first places the river came over. At 11:02 McCormick called headquarters on his walkie-talkie. "The wall's broken! There's no way to stop it. Some of the volunteers are still here – we've got to get them out. Tell the General to blow the sirens!"

For reasons that may never be known, the sirens weren't blown until 11:16, fourteen minutes after McCormick alerted Civil Defense about the breach.[1] During that time the river topped the wall at several other places and began pouring into Wilkes-Barre. A photographer for the Times-Leader captured the scene as hundreds of young volunteers fled the watery onslaught, one of the iconic images of the disaster.

The sandbag army couldn't keep the river from coming over the walls, but it did buy time, crucial hours when the people of Wilkes-Barre and neighboring communities could pack and leave. "The kids were tremendous, thousands of them shoveling mounds of sand. They really jeopardized their lives, but they held the damn river back for three hours and gave people a chance to get out."[2]

Sand-baggers flee through the mud as the Susquehanna River breaks over Wilkes-Barre's flood walls (Courtesy Wilkes-Barre Times-Leader)

Only Bill Shock and a handful of others remained at Ross Street. Caught up in their mission, they were oblivious to the increasing danger. "Just one more," they shouted, over and over, as they valiantly but vainly tossed bags of sand into the widening gap at the wall. And then, the screech of the sirens, seven short bursts. "There it is! Let's go!"

Shock scrambled down from what was left of the sand pile. Rushing water tugged at his legs. He cursed as he stepped on a discarded shovel and lost his balance. For the first time, he was frightened. "Hey guys!" But the others were too far ahead. Another segment of the sandbag wall collapsed, followed by another fire-hose of flood water. Shock could no longer move against rushing water suddenly up to his waist. He fell again, managed to stand, then slipped and went down. The current carried him to the payloader abandoned by his friend Howie, and Shock grabbed hold. Maybe he could hang on until someone came to help. But no one would come. Shock slipped beneath the loader as the water rose higher, his legs entangled in the steering rods. There was no escape, and he couldn't keep his head above the flood. Bill Shock took his last breath as his lungs filled with muddy water. His body, still in his yellow slicker, would not be found for six days, carried by the flood to lodge in a backyard blocks away from the river. Agnes brought death to Wilkes-Barre.[3]

* * *

Police, fire, and Public Works radioed in to Civil Defense's temporary headquarters at the sheriff's office as the Susquehanna breached the dikes all along Wilkes-Barre's river-front. *It's coming over at Charles Street! ... The Market Street Bridge is flooded over! ... Water is into the library at Kings College!*

WILK was the only radio station in Wilkes-Barre still on the air. Its news director David DeCosmo, pressed into service as Civil Defense spokesman, broadcast Townend's order at 11:30. *Get out! Get out now! If you have not evacuated Wilkes-Barre, get out NOW!*

Souchik went to the window at the end of the hallway and

looked out at the brown water now swirling around the courthouse. A squad of National Guard was building sandbag barriers around the doors of building, but it seemed too late. The little shed housing his river gauge was half submerged, and there was no way he could reach it. Beyond the gauge house, logs and other debris piled against the North Street Bridge. Souchik hurried back to his temporary headquarters. "General, we can't stay here. We'll be stranded. And if it gets in the building the generator will flood out."

Townend turned to communications director Pissot. "Bob, get Fred Kepner at the school district. I think they have space in their admin building we can use. And get the 109th to send a couple of trucks down here." Then he raised his voice, addressing everyone in the room. "We've got to move. Again!"

* * *

On Friday afternoon a helicopter flew over Wilkes-Barre, carrying an incredulous news crew from CBS-TV. *Unbelievable! Wilkes-Barre, completely inundated. You can't even see a road, a side road, an alley, a back street or a back yard. The water is cresting over the first floor of many of the homes. All of it is muddy Susquehanna!*[4] The images flashed on television screens across the country, and the world learned of the devastation of the Wyoming Valley,

The rampaging river crushed forty percent of Wilkes-Barre and its neighboring communities, twelve miles along the river on both sides. The flood raged twelve feet deep at Public Square, destroying the offices and shops surrounding it. At the venerable Sterling Hotel it smashed through the glass doors of the ornate entryway and gutted the lobby, dining rooms, and Lenchen Townend's little Wide Awake Bookshop, leaving a sea of mud and moldy carpet. Along Riverside, Charles, Academy, Market, Ross, and every other street within a half mile of the river, homes and businesses, thousands of them, ripped from their foundations, buckled under the strain of the surging flood, or were left gutted. At Kingston, only twenty of the city's six thousand homes remained undamaged.

Rescuers guided small boats on Wilkes-Barre's flooded streets as fires raged uncontrolled (Courtesy Wilkes-Barre Times-Leader)

A block of warehouses caught fire a quarter mile from the river. Firemen couldn't get their trucks through the nine feet of water surging up Northampton Street. "There's a big fire raging, but they can't do anything about it!"[5] The towering plume of black smoke could be seen for miles as the entire block of buildings burned to the waterline, leaving only gutted brick shells.

* * *

Helicopters pilots who learned to skim over the jungles of Vietnam flew in the front line of flood response. Now, as pilots for the Navy, Coast Guard, and state police, they rescued hundreds of victims and delivered crucial medical supplies. At Falls Township, just upriver from Wilkes-Barre, a family of four was spotted on the roof of a house that had lifted from its foundation and floated down the Susquehanna. State police pilot Gordon Walchowski hovered over the stranded family until he plucked them all to safety, before the house overturned or smashed into a bridge.[6]

* * *

Bill Seiwell, a 27-year-old mechanic, heard Townend's call for "anyone with a boat" to help with rescues.[7] He and a companion had already saved eight people from flooded homes in downtown Wilkes-Barre. Now, as he guided their flat-bottomed bass boat through nine feet of water south of Public Square, there were cries from the left. Two women huddled on the front porch roof of a double block, waving towels to attract attention. Seiwell guided his little boat to the frantic women, fighting whirlpools formed as the current raced around the sides of the old brick home.

The surface of the water rushed by only inches below the edge of the roof. With one foot in the boat and his other foot on the eave, Seiwell reached to the terrified women while his partner struggled to keep the little boat still in the swirling current. The women stepped gingerly, inch-by-inch, toward Seiwell's outstretched arms. He shifted his weight to the roof to extend his reach as the women crept closer. But the old shingles were slick with moss, and Seiwell slipped back. Water splashed over the gunwale as Seiwell's weight shifted. "Bill, watch out!"

Seiwell jumped too far, and the little boat pitched sideways. As the women scrambled back onto the roof the two boaters were thrown into the brown water. Seiwell's companion clung to the upturned boat until it lodged against some trees a block away, but Seiwell never came up. His body was found by a state police scuba diver the next day, stuck beneath a loading platform at the nearby Planters Peanut factory.[8]

* * *

Families were separated in the confusion of the city-wide rush to safety. Roz Smulowitz and her 18-year-old son Jacob joined the sandbagging crew across the street from their Riverside Drive home. "We were filling pillowcases, trash bags, anything. Hundreds of people were working their guts out. But the river kept coming, and the sirens blew. Everyone ran. I was separated from my son, and began shouting for him. It was two days of hell before we found him. Thank God he was alright."[9] Days later, when Smulowitz made her way back to her home, she found it ripped from its foundation, torn in half, the remnant of it leaning against a tilted phone pole half a block away. "All we salvaged from that house was our love. Everything else, everything, was gone."

At Pocono Downs race track, set up as a distribution center for food, clothes and supplies, an elderly woman sorted boxes of donated food, tears streaming down her cheeks, "I can't find my daughter and her family, I don't know what to do except pray".[10]

All across the valley communications were down. The phones didn't work, and in much of the city there was nowhere, and no way, to deliver mail. Emergency radio was overwhelmed with official business. All of the bridges across the Susquehanna were destroyed or closed, and a dusk-to-dawn curfew enforced by the National Guard restricted movement even on foot.

To help reunite families, and to assure refugees their loved ones were safe, WILK alphabetically broadcast the names of every person at all eighty-one of the public shelters. *The following individuals have checked into the B'nai Brith Home on Northampton Street: Jeffrey Aaronson, Laura Abbenante, Daniel Abbot ....* It would take two hours to get through the entire list. And then, start over.

***

It wasn't only people who were rescued. Many residents left their homes with only the clothes on their backs, not imagining how high the flood would rise or how long they would be gone. Often, their pets were left behind.

Early Friday morning a woman came into the Luzerne County SPCA shelter in Plains with her little Boston terrier on a leash. Could the shelter keep the dog while the family stayed with relatives?

Attendant Katey Taylor, twenty years old, took the leash. Yes, they would keep the dog. A couple of other people had already dropped off their cats.

The woman relaxed. "Thank you! We were worried we'd have to leave him." She hesitated at the door as she was leaving. "You know, the people next door to us left two dogs chained in the yard. I hope they'll be okay."

Taylor got the address. They would have to check this out.

Half an hour later, Taylor and fellow SPCA driver Maureen McGlynn were on their way to Charles Street to check on the abandoned dogs. Their radio was tuned to WILK, on the air with constant flood warnings from Townend and spokesman David DeCosmo. The women saw the legions of volunteers working at the wall, and a steady stream of residents walking or driving away from the river. "Maureen, a lot of these people are leaving their animals behind. We have to do something!"

Over the next three days, as the river surged over the wall and into Wilkes-Barre's neighborhoods, Taylor and McGlynn were on a mission to save the animals.[11] Their van was useless in the flooded streets, so they borrowed a boat from the fire department. Working twelve hours a day they answered calls and tips about stranded pets. *Two cats on a roof in Plains ... A dog is trapped in a house on Madison ... A family evacuated to the high school left their pets at home.* By the time the flood receded on Monday the women rescued more than a hundred frightened, quivering animals, often breaking windows to get into abandoned houses.

Animals other than family pets were in trouble too. Countryside farmers sloshed through chest-deep floodwater to release their cattle, horses, and other livestock into open fields, hoping to round them up after the storm. In the rail yard at Kingston, boxcars carrying a traveling circus were knocked on their sides. SPCA volunteers safely removed all the show's animals – horses, llamas, two camels, a bear, and several smaller creatures – and secured them on dry land. The owner of a roadside petting zoo near Covington cut the wires around his pens to free his deer, elk, sheep and goats into the nearby forests. When the flood receded all

Many pets and other animals were rescued by the SPCA, but thousands more did not survive the flood (Courtesy Bryson Leidich Photography)

of his creatures, soggy and hungry, but for two peacocks, returned on their own.[12]

Despite the heroic work of the SPCA, farmers, pet owners and volunteers, Agnes took the lives of thousands of animals. Hundreds of pigs, cattle and horses did not escape, their carcasses swept away in the flood to lodge on the riverbanks.[13] Poultry houses were especially hard hit, and an untold number of chickens and turkeys swept away. Hundreds of dogs, cats and other household pets left to fend for themselves, some chained in their yards, drowned in the flood. And, Bill Schwab's parakeet, left behind when the family fled to his sister's home, was found dead when Schwab returned five days later, its cage six feet above the floor no refuge when the water rose to the ceiling.

* * *

Wilkes-Barre and its sister cities along the Susquehanna River in Pennsylvania's Wyoming Valley suffered more damages, by far, than any other community caught in the maelstrom of Agnes. The Corps of Engineers pegged property destruction in Luzerne County at $1.1 billion, approaching "the greatest cost ever experienced as the result of a single natural disaster".[14]

Yet amidst the flood, the fires, the destruction of homes, businesses and infrastructure, the breakdown of communications, and the hasty evacuation of a hundred thousand residents, Wilkes-Barre and the surrounding Wyoming Valley suffered only five fatalities during the Agnes flood.[15] It seemed almost a miracle. But hydrologist Gwinner warned the city, for a few precious hours volunteer sandbaggers staved the river off, and residents heeded General Townend's dire pleas to get to safety.

Pennsylvania's Governor Shapp sent a personal letter of gratitude to the hydrologists at the River Forecast Center. Gwinner's "recommendation that everyone behind the dikes be evacuated," the governor wrote, "was responsible for saving countless human lives."[16]

LOGS

Friday, June 23, 1972
Forty Fort, Pennsylvania

The village of Forty Fort, on the west side of the Susquehanna River, stands a mile upstream from Wilkes-Barre. During the American Revolution, the settlers of Forty Fort assembled a ramshackle militia to defend against marauding British troops and their Iroquois allies. Alas, the farmers were no match for the seasoned Redcoats. On July 3, 1778, the militia was routed in what became known as the Wyoming Massacre. Three hundred of the defenders fell during the battle or were captured and murdered over the next few days. *Remember Wyoming* became a battle cry of the Revolution, and the sacrifice of the Forty Fort militia is indeed remembered in the name of the 44th state, Wyoming. When they died years later, survivors of the battle and families of the fallen were interred in the big Forty Fort Community Cemetery along the banks of the Susquehanna.

* * *

On Friday morning Joe Ostrowski,[1] on summer vacation after his junior year at Penn State, stood in with drizzle two friends at the Forty Fort cemetery gate. An engineering student, Ostrowski just completed an introductory class in river hydrology. He'd been following reports of the rising flood with rapt attention, heard General Townend's warnings, and, now that the worst of the rain was over, wanted to see the river with his own eyes.

An earthen levee and six-foot steel wall, as at Wilkes-Barre, had been built along the river at the far side of the cemetery to protect Forty Fort from floods. Ostrowski and his friends walked briskly toward the levee along a gravel path amid centuries-old gravestones. When they reached the levee they scrambled to the top and chinned up the steel wall to peer over. The muddy water of the Susquehanna swirled inches below their noses. "Holy crap! It's going to come over!"

One of Ostrowski's friends stiffened and pointed up river. "Look! It already is!" Some years before, a mine tunnel under the river collapsed, sagging the levee above it and tilting the steel flood wall atop the levee inward. The boys watched, mesmerized, as backwash from water pouring over the sag eroded the levee and widened the breach. A giant whirlpool formed where the river gushed over the wall and gouged at the soil in the cemetery. Frozen in shock, Joe and his friends gaped at what looked like big boxes popping out of the ground. "Oh my God! The coffins are coming up!"

"We did the only logical thing we could do," Ostrowski said later. "Run!"

The whirlpool grew stronger even as Ostrowski and his friends fled. The force of the water toppled tombstones, family vaults,

The old Forty Fort Community Cemetery was gouged open by the flood, exhuming 2,500 burials (Forty Fort Cemetery Association)

fences and trees. A four-acre section, nearly a third of the cemetery's area, disappeared into a pit twenty feet deep. 2,500 graves were disinterred, many of them washed down the river or reburied in the sand and mud left behind.

A mile upriver, Joe Shaver[2] and his neighbors stood on the levee watching the churning Susquehanna. Shaver, a funeral director by trade, served as deputy coroner for Luzerne County. What Shaver and his group saw made no sense. "The river is going the wrong way!" The flow around big Mononcanock Island formed a gigantic eddy to push the current upstream along the west bank. "Look! Are those logs floating in it?"

Shaver peered at the dark "logs" bobbing in the huge waves. When he saw that the logs had arms and legs he recognized them as corpses, maybe ten of them. He knew they could only have come from burials at the Forty Fort cemetery.

Shaver and chief coroner Dr. George Hudock set up a temporary morgue at the historic Swetland Homestead, a block from Forty Fort's Wyoming Battle Monument, to handle the disinterred remains. It wasn't the coroner's usual role to deal with cadavers already buried, but there was no one else qualified for the job.

The Army sent troops from its mortuary unit to help sort and identify the bodies, and the Marine Corps sent men with equipment to move the heavy concrete vaults dislodged from the cemetery. Stoic volunteers, piloted in small Coast Guard boats, helped with the search. Young Ostrowski was offered a job bagging remains, but declined. "I didn't want to have nightmares for the rest of my life."

On Saturday the volunteers brought in ninety cadavers, on Sunday a hundred and fifty, on Monday five hundred. Most were found in and around the old cemetery, the Little League fields nearby, or the backyards of neighboring residents. Some were still in coffins but many were not. Bodies were found in backyards, in garages, tangled in trees and fences. At first, many of the corpses were thought to be flood victims. Some were well preserved, while the older ones were just bones; most were somewhere in between. There wasn't nearly enough room in the temporary morgue to store them all, so the bodies were bagged and stacked along Wyoming Avenue.

Families of the deceased, their grim faces masked to ward off the stench, trooped into Swetland Homestead. "Have you found my grandmother? My brother?" But the flood washed away most of the cemetery's records. Of the 2,500 burials that had been disinterred only half were found, the rest of them lost forever in the flood. In the end, Shaver could identify only thirty-six.

In the week it took to collect the cemetery remains, Shaver, Dr. Hudock and their volunteers barely slept. Reeking from mud and decay, they took their meals and stayed at nearby Wyoming Methodist Church. When Shaver finally returned home he had to step over a dozen strangers sleeping in his living room – his wife had opened their door to refugees from the flood, feeding and giving dry clothes to people who lost everything.

Years later, the federal government wanted to enlist Shaver as an expert on recovery from cemetery floods. "Once was enough," he replied. "I never want to go through that again."[3]

Many of the remains recovered from the Forty Fort cemetery were re-buried at Memorial Shrine Cemetery at Dallas, a couple miles up the hill. The eroded land at Forty Fort was filled and replanted, but the hallowed ground could never again be used for burials. In 1979 the Cemetery Association erected an eight-foot-tall granite obelisk remembering those lost:

On the afternoon of Friday, June 23, 1972
The Susquehanna River
Swollen by flood waters of unprecedented height
Broke through the dike at a point 120 yards south of this site
The swirling water gouged a four-acre chasm
Out of the heart of the cemetery
Displacing approximately 2500 burials
This Park is dedicated
To the memory of those whose graves vanished
In that singular catastrophe

WE LOVE YOU

Thursday, June 22, 1972
Lewisburg, Pennsylvania

The headwaters of Pine Creek sprout in the Twin Tiers of New York and Pennsylvania, right in the bullseye of extreme rainfall from Agnes. The little river hurtles south though an 18-mile-long cleft in the Endless Mountains known as the "Grand Canyon of Pennsylvania," then joins the West Branch of the Susquehanna near the little city of Jersey Shore – so named by a pioneer family from New Jersey.

In the Nineteenth Century, Pine Creek and the West Branch served as highways for the logging industry. Burly raftsmen piloted huge rafts of timber from the forested hillsides downriver to market, and rest stops along the way grew into the little cities of Lock Haven, Renovo, Williamsport, Montoursville, Lewisburg and Milton. The rafters typically floated their timber down with high water of the spring thaw. But they never saw anything like Agnes.

On Wednesday, eight inches of rain over the watershed of the West Branch flash-flooded tributary creeks into little villages. Bridges over normally benign little creeks washed out, and mudslides roared down steep slopes to block roads.[1] The rain was thought to have ended Wednesday evening, and the River Forecast Center predicted a modest 17-foot crest of the West Branch at Williamsport by Friday.[2]

As occurred on the other rivers emanating from the epicenter of extreme rain in the Twin Tiers, an unforeseen surge of floodwaters rushed down the West Branch Wednesday night and into Thursday, swamping the little cities along the way. Muddy water roared twelve feet deep through the streets of Lock Haven, and three of the city's four bridges washed away.[3] The river rose to a record 35 feet on the Williamsport gauge, more than twice as high as predicted by the

Forecast Center only two days before; fortunately, new dikes kept most of downtown Williamsport dry.

At Lewisburg, Public Safety Director Gordon Hufnagle guided his little rescue boat toward Joseph and Agnes Murphy, stranded in rising waters near the campus of Bucknell University. Moments after the Murphys climbed aboard the little craft it capsized, and the three were pitched into the whirling water. Joseph Murphy hung onto a nearby tree and was rescued, while Hufnagle and Agnes Murphy tried to cling to the overturned boat. "The last time I saw them, the current whipped the boat under the railroad underpass," a witness reported.[4] The body of Gordon Hufnagle was found a block away five days later; Agnes Murphy would not be found for two weeks, five miles down the West Branch.[5]

* * *

Thursday, June 22, 1972
Sunbury, Pennsylvania

Mildred Gold might have looked over the beautiful Susquehanna River from her home on Sunbury's Front Street if the flood wall didn't block her view. As it was, she had no river view at all, just the long grey slab of concrete, eight feet high, running along the opposite side of the street. At least the bluffs of Shikellamy Overlook peeked over the top of the wall, a hint of the picturesque landscape beyond.

Sunbury's flood wall, as at Wilkes-Barre, was built after the 1936 St. Patrick's Day disaster to protect the city from the Susquehanna. It was completed in 1951, long before Gold and her husband Nathan moved into their big brick home. Sure, they heard that back in '36 the water came into the basement, but since they'd lived there the river never really threatened. Lately some of Gold's neighbors complained about the wall to the city council and planning board:[6] tear it down and make a riverfront park! This was the more affluent part of town, and many believed the ugly wall cut into their property value.

Sunbury, forty miles north of Harrisburg with a population of about 13,000, was within the zone of Agnes' heaviest rain Wednesday and Thursday. Shamokin Creek, running along the eastern and southern edges of town, flash-flooded to 6,000 cfs, almost double its previous record.[7] Muddy water sloshed into the city water plant, and the county animal shelter was swamped; fortunately, all of the dogs and cats were moved to safely.[8]

Mildred's husband Nathan Gold, a dentist, arrived home early Thursday afternoon. The main roads into town were cut off by flooding streams, and the mayor ordered all businesses to close for the day.

"We'll be fine!" Mildred was unconcerned. "I've been watching the news. The rain is supposed to stop by tonight. They've been getting floods down around Harrisburg, but they always do, don't they?" She paused, then added, "I also heard they deflated the fabri-dam." Sunbury had a unique inflatable dam spanning the river to form a lake for boating, water-skiing and fishing.[9]

Nathan looked puzzled. "They only let that down for a serious flood. That's not going to happen ... is it?"

"All I know is the radio keeps saying that flood stage is twenty-four feet," Mildred harrumphed, "and the river is only at eleven feet, whatever that means. But I know that damn wall is thirty-five feet. Which proves we really don't need it."

Sunbury is at a crucial point on the river, where the West Branch meets the Susquehanna coming down from Wilkes-Barre. Forecasting the river here is complicated, as the amount and timing of the flow from both branches must be accounted for.

The crest of the West Branch came first, after rampaging through Lock Haven, Jersey Shore, South Williamsport, Milton, and Lewisburg Thursday afternoon. But the West Branch alone, although at its all-time record flood, didn't pose a threat to Sunbury. The crest was still ten feet below the top of the wall.

On Friday morning the Golds sat in their breakfast nook sipping coffee, listening to WHP radio from Harrisburg. The news reporter said the flood on the Susquehanna upriver had become unexpectedly worse overnight. No sooner had they heard O.D.

White's 7:15 interview ... *record stages along the entire river* ... when there was a loud knock at the front door. Mildred answered it to find a Sunbury police officer in a grey slicker, rainwater dripping from his service cap.

"Mrs. Gold?" The policeman glanced at his clipboard. "Everyone along the river, from Front Street back to Fourth, has to evacuate, in case the flood breaks through, or even comes over the wall."

Mildred looked past the policeman to the wall across the street. It was still ugly, but now seemed truly ominous.

"You need to leave by noon. Mayor's orders. Do you have somewhere to go?"

Yes, Nathan's partner lived up the hill, they could go there for the night.

The policeman continued. "Things could get bad if it comes over the wall. Take your important papers and medicines. And you might want to move some things upstairs." He put a check by their name on his clipboard. "I have to get along. Good luck!"

Mildred and Nathan Gold, along with two thousand other residents of Sunbury, did as the mayor ordered and fled the Susquehanna's reach.

By Friday afternoon the flood exceeded the 1936 record, although still several feet below the top of the wall. Pressure on the wall was tremendous as the river continued to rise through the night. It wasn't built to take a prolonged assault like this.[10] Water bubbled up from beneath, leaked through at the seams, and splashed over the top. City crews and volunteers sandbagged low spots, plugged the leaks, and reinforced the bridge gates with timbers.

The Golds were glued to the television in the safety of Nathan's partner's home. They watched in horror at images of the disaster unfolding at Wilkes-Barre upriver, and at Harrisburg downriver. But around Sunbury – heard on local radio, in the Daily Item, neighbor-to-neighbor – *the wall is holding*. Still, the weatherman said that the river wouldn't crest until Saturday. The Golds went to bed not knowing if their home on Front Street would survive the night.

On Saturday morning the RFC issued a statement. *Severe flooding continues throughout the Susquehanna basin. The most critical spot is Sunbury, where the water is at the top of the dike, and rising.*[11] Hydrologist Michael Mark explained, "It could be just inches. We don't know if the dike will be topped."[12] Even inches over the wall would have inundated Front Street and most of downtown. The wall was already "leaking like a sieve,"[13] and in places the river splashed over. All eyes in Sunbury were on its wall.

The river crested at Sunbury, 35.8 feet, at noon Saturday. This was half a foot above the "official" height of the wall, but a margin of error had been factored into its construction. The wall held, and Sunbury was spared from disaster. If the crests of the West and main branches of the Susquehanna came at the same time, if the river ran only two inches higher, the Gold's home and hundreds of others would have filled to the second floors with the raging, muddy river.

The river crested at the very top of Sunbury's flood wall. Homes and businesses throughout Sunbury were spared destruction (U.S. Army Corps of Engineers, The Corps Responds)

The city's Civil Defense director summed it up: "If it weren't for the dike, Sunbury wouldn't be here today."[14]

Mildred Gold stepped in through her front door on Tuesday. The river was back down to twenty-four feet, still above flood stage but deemed no further danger. There was no damage to the Gold home, and everything remained as they left it. Nathan laughed that he hadn't mowed the lawn, and with all the rain it looked like a jungle. Still, a week went by before the city's water plant was repaired and service turned back on. Some roads and bridges were closed for weeks, but Sunbury survived the flood mostly unscathed.

The next Thursday morning Mildred Gold looked from her front window and smiled. She called her husband to join her. Someone – in what at another time would be called vandalism – had scrawled huge black letters on the dirty concrete slab that faced their home.

Nathan put his arm around his wife. "You know, it's not so ugly anymore."

(U.S. Army Corps of Engineers, The Corps Responds)

GASLIGHTS

Friday, June 23 – Saturday, June 24, 1972
Harrisburg, Pennsylvania

The "Old Shaky" Walnut Street bridge lived up to its nickname as Gwinner gingerly walked to the middle. The usual gauge for Harrisburg, at the foot of Nagle Street, flooded out the night before, so Gwinner was on his way to the alternate wire weight gauge on the bridge. The gauge was mounted halfway to City Island, on the upstream side of the bridge so the operator could see dangerous flotsam approaching. Bits of houses and barns, uprooted trees, overturned boats, and dead livestock floated down the churning river. When larger pieces hit the bridge the entire structure jolted and threatened to crumble from its piers.

Gwinner unspooled the weight from its housing until it touched the river surface, now just a few feet below the rail. He squinted at the display to read 28.85 feet – a tad higher than the 1936 record.

Instead of going back to his office, Gwinner walked from the bridge to Civil Defense headquarters at the Dauphin County courthouse a block away, where he would phone his reading in to his colleagues at the River Forecast Center. He wanted to take another measure at 8:00, and it was more convenient to stay close to the bridge. Besides, with no sleep in the last twenty-four hours, he was eager for the fresh coffee and donuts he knew would be on hand at the Civil Defense office.

The power throughout Harrisburg blinked off moments after Gwinner reported his new-record reading. Except for a battery-powered "exit" sign, the basement office of the Civil Defense office went pitch dark. As at the RFC four blocks away, the Civil Defense staff froze, not knowing what would come next.

The darkness didn't last long. There was a rumble as the courthouse's diesel generator kicked on, and the lights and equipment flickered back to life. The Civil Defense staff exhaled, and pandemonium picked up where it left off. Gwinner called the RFC to find out if the power was out there too, but gave up after the phone rang for two minutes with no answer. The RFC staff had not yet figured out that they had to pick up their silent phones to determine if anyone was on the line.

A few minutes before 8 a.m., Gwinner, bolstered by the coffee, donned his yellow rain jacket and pork-pie hat and headed back to the bridge to take another reading. Higher now, the river sloshed up through the bridge's iron grate deck. Gwinner's shoes were soaked with muddy water by the time he reached the gauge. The old iron bridge creaked and groaned as ever more flotsam jammed against its upstream side. Gwinner knew this wasn't safe, but also knew his measures were crucial. He quickly took the reading: 30.65 feet, almost a foot higher than an hour before.

Just as Gwinner closed the gauge box a big oval form bobbed down the roiling Susquehanna. He recognized it as a propane tank, followed by two more, unleashed from somewhere upriver. Gwinner gripped the rail, his knuckles white, afraid the tanks might spark and explode when they hit the metal bridge. One by one the tanks crashed into Old Shaky. Two of them lodged in the debris piled against the upstream side, while the third screeched beneath and wrenched the handrail and walkway off the downstream side of the bridge.

A city policeman at the entrance to the bridge waved frantically as Gwinner sloshed back toward safety. "Get off there! The bridge is gonna go!"

Gwinner identified himself and explained his mission. The policemen relaxed, a little, and gestured to his patrol car. "Come on, I'll give you a lift back to the Fed building." Exhausted, soaked, and shaken, Gwinner gratefully accepted.

The Walnut Street "Old Shaky" bridge survived Agnes, but was so battered and unstable it was never again safe for traffic. Popularly called the "People's Bridge", it remains to this day a pedestrian-only crossing to City Island.

* * *

The walk up eleven flights of stairs at the Federal Building left Gwinner panting. As he shucked his rain jacket he spotted O.D. White and hydrologist Don Close near a window, sorting through a stack of old charts. White turned to him. "You look like a drowned rat, Mike. You've done enough. Go home and get some rest."

"I will, soon. But look, the Walnut Street gauge is done. Water is over the bridge deck and there's a ton of debris. It was shaking so hard I could barely stand."

White nodded. "All the bridges except I-83 are closed." Patsy Quigley had a battery-powered radio she listened to on her lunch break, now the RFC's only source of public news. Nevertheless, the RFC needed that river data, every hour now. He turned to Close. "Don, isn't there's an old portable wire-weight in the equipment closet? See if we can use it at Market Street."

The Market Street Bridge, a quarter mile downriver from Walnut Street, was built with a core of solid concrete and armored by granite blocks to defend against the river. The western end of the bridge was submerged in the flood, but the eastern end downtown stood proud. Hydrologists Close and Mark installed and calibrated the portable gauge on the sturdy stone rail by 11:00 a.m. and took their first reading: 31.4 feet, almost a foot higher than Gwinner's reading at Old Shaky three hours earlier.

As the hydrologists stepped off the bridge onto Front Street they were confronted by two fresh-faced National Guardsmen, M-16s slung over their shoulders. "Stop! You're in violation of curfew."

Close and Mark explained they had the mayor's permission to install a gauge on the bridge, and were on their way back to the Federal Building.

The young guardsmen never heard of the River Forecast Center, and had no instructions about how to handle this situation. Their orders were to bring would-be looters to the armory, but this seemed legitimate, maybe important. "I have to check this out. Stay right here." The young guardsman thumbed his walkie-talkie to call his superior.

Twenty minutes went by before the guardsmen received a reply from their commander. "These men are cleared. Escort them to the Federal Building without delay."

As grey light filtered through the windows at the RFC the hydrologists pulled out the paper charts and tables for all stations on the river, searched through their drawers for their old slide rules, and manually entered the latest precipitation and stage data into the forecast algorithms. This was how river forecasting was done twenty years before, and how they had to do it again.

White re-ordered the office tasks to accommodate their circumstances. One of the technicians marched down to Market Street every hour to read the temporary gauge, armed with an order of passage signed by the mayor. While White's hydrologists calculated the forecasts, sketched new charts, verified the prior predictions, and prepared bulletins, other staff were assigned a silent phone to pick up every few minutes to answer calls. There was a standing call to Civil Defense at half past every hour to exchange data and issue updates, while the most urgent communications were hand-carried. The RFC routed some of its communications through the Weather Service office at Williamsport, where updated forecasts were disseminated to regional weather offices; in turn, Williamsport collected data relayed from Binghamton, Trenton and Washington. The RFC operation was held together with baling wire and duct tape, but as slow and cumbersome as the process was, it worked.

Gwinner, on duty for twenty-five hours, finally went home at 3 p.m across the still-open I-83 bridge. He returned at 7:00 the next morning for another thirteen-hour stint.

* * *

No one knew when the power might come back on. As the river kept rising and crept ever more deeply into downtown Harrisburg, White didn't think it would be anytime soon. At 5 p.m. he told the staff he was going home for short time, then returned a little after 7:00. The hydrologists and staff had consumed little more than coffee all day, and didn't realize how hungry they were until White plunked two buckets of fried chicken on the conference table.

Two of the staff walked down, and up, the twelve flights of stairs to the basement parking area to retrieve two Coleman lanterns and an air mattress from White's car. He also booked rooms at the venerable Penn Harris Hotel across the street, where his staff could, in rotation, catch a few winks. For many of them, the roads to their homes were closed.

On through Friday and into Saturday morning, faced with the most widespread flooding in the history of the United States, and the most daunting and critical task of their lives, O.D White and the hydrologists at Harrisburg worked through the night by gas light, with pencils, old paper charts, and slide rules. Their modern equipment – radios, teletypes, typewriters, and most crucially their IBM computer, were useless. It went unsaid, but everyone at the RFC was well aware that their forecasts literally meant life or death for hundreds of thousands of people who lived near the mighty, rampaging Susquehanna River.

* * *

Grey dawn broke Saturday as low clouds swept across the skies from the west. The counter-clockwise rotation of Agnes pushed on to the north, and there would be glimpses of blue sky by late afternoon. Yet even as the clouds parted the big river kept rising, fed by the surge still coming from upstate. The RFC's overnight calculations predicted crests at Wilkes-Barre about noon, at Harrisburg a few hours later.

White was proud of his hydrologists. They worked with little rest, handicapped by sparse data and failure of their equipment. Their forecasts remained the basis of crucial decisions and actions made by public agencies, first responders, thousands of businesses, and a million residents.

White had to get his computer and other equipment working again, and called state Civil Defense chairman John Gristell for help. Gristell ordered the National Guard to deliver a powerful generator to the Federal Building, highest priority. On Saturday afternoon Gwinner went to the roof with two Guard technicians to receive the generator. They heard the whomp of rotor blades before a heavy Chinook helicopter, with the generator dangling beneath it, swung into view. Two hours later, the technicians, a lineman from Pennsylvania Electric, and the superintendent of the Federal Building had the generator up and running. The RFC's lights flickered on, and the first sound of a ringing telephone came moments later. The hydrologists cheered, but barely paused in their work; within minutes, Quigley was punching data compiled by hand that morning into the IBM 1130.

* * *

A wall of water is coming down the river! A buzz went through the RFC Saturday night.

White was skeptical. "Who says so?"

Hydrologist Nick Pavick had taken the call: a citizen at Highspire reported a wave three feet high headed toward Three Mile Island. The big power plant under construction five miles downriver from Harrisburg was two years away from completion, and there were no nuclear materials on site. But the protective flood walls at the island's perimeter were not yet complete, and the reported wave could overrun the unfinished plant and set construction back for years.

A tsunami-like wall of water can come after the sudden failure of a dam. White knew there were no dams for miles above Harrisburg. Perhaps a debris dam at the I-83 bridge or Pennsylvania Turnpike had broken. Trees and other flotsam accumulated at bridges the entire length of the river, temporarily damming the flow behind them. The river could surge through if the dam of debris were to break, as happened at Corning. White had to verify the frightening report. "Okay Nick, see if you can alert the manager at the plant. We can't do anything about it, but they need to know."

Three Mile Island's plant manager called back an hour later. The report turned out to be just one of many rumors during the "fog of flood". There was no tsunami-like wall of water, but Susquehanna River at the power plant was above its record crest, and if it got any higher would wash over the site.

* * *

At noon Friday, hydrologist Feese called to White across the room. "I've got Paul English at Conowingo on line 3. Can you pick up?" English, a civilian employee of the Army Corps of Engineers, was superintendent of the giant hydroelectric dam sixty river miles below Harrisburg. "He needs to know how many of his spill gates to open."

Debris piles up at Harrisburg's "Old Shaky" Walnut Street Bridge, where Gwinner ventured out to read the wire-weight gauge. The black-cube Federal Building stands in the background (Courtesy Bryson Leidich Photography)

DYNAMITE

Friday, June 23 - Saturday, June 24, 1972
Port Deposit, Maryland

The 27,000 square-mile watershed of the Susquehanna River might be thought of as a gigantic funnel. All the water that drains down the North Branch, West Branch, Chemung, Juniata, and hundreds of smaller tributaries flows through the narrow point of that funnel at Conowingo, just before discharging into Chesapeake Bay. The power of the great river at this point is enormous, power that long begged to be harnessed.

Five thousand construction workers, living in a nearby tent city, toiled more than two years to build the great Conowingo Dam.

Dramatic artistic rendition of Conowingo Dam (1950s postcard)

When completed in 1928 it was one of the biggest dams in the United States, a dramatic Art Deco structure more than a hundred feet high and almost a mile across, holding back the Susquehanna River for hydroelectric power. Although surpassed in the 1930s by dams on the Tennessee, Columbia and Colorado Rivers, Conowingo remains the largest hydroelectric dam in America's Atlantic watershed and a key part of the regional power grid.

* * *

Dam superintendent Paul English slammed the phone down. It was 10 a.m. Friday, June 23. He had no idea the power was out in Harrisburg and the Forecast Center's phones hobbled. The forecast relayed via teletype from Civil Defense at Harrisburg was two hours old, and his reservoir was already higher than predicted. He needed to talk directly to O.D. White for an emergency update. When English finally got through two hours later, all White could promise was a new forecast later in the afternoon.

* * *

Dave Goodland, hired only six months before, couldn't believe his luck landing a job as a "dam operator". While the engineers and technicians in the control room managed the dam and power plant, Goodland and a dozen other operators did the hands-on work. The union job paid well and always kept his interest. Someday, Goodland figured, he'd work his way up to the control room.

Back in March, Goodland's foreman showed him how to raise the huge spill gates that topped the dam. Then, the river was running high with late winter snow melt, more flow than could be handled by the penstocks through the power plant. To keep the reservoir from getting too high, the superintendent ordered his operators to open one of the fifty-three spill gates atop the dam to release the excess flow. Under his foreman's watchful eye, Goodland guided the sixty-ton crane along the rail that ran the length of the dam. He stopped over Gate #6 and levered the controls to slowly hoist the massive steel gate. The river poured through the open gate and plunged to the rocks below. His boss approved. "You've got it Dave. You know, they say that back in 1936 the river got so high they opened all of the gates for a couple of hours. Man, that would be like Niagara Falls!"

* * *

Goodland arrived at the dam Friday morning, June 23, to start his eight-hour shift. The night crew had already opened two of the spill gates, and the foreman sent him out in the drizzle to see if debris had collected around them. The reservoir, dark brown with silt, looked like a foaming cauldron of coffee, with bits of houses and barns, overturned boats, trash, and uncountable broken trees swirling through it.

When Goodland saw how high the reservoir was rising, and how much water poured through the two open gates, he wasn't surprised when his foreman crackled in over the walkie-talkie. "Goodland, get over to the north end and open ten gates. You copy?" The transmission was filled with static.

Goodland toggled the "send" button. "Yes sir, open gate number ten. Got it."

The reply was immediate. "NO! OPEN TEN GATES!"[1]

Goodland had never seen more than two opened at once. It took him four hours to guide the crane along the rails to lift five at the north end, then five more at the south, roughly balancing the spill. Still, it looked to him as if opening gates had no effect. The surface of the reservoir kept rising.

* * *

In the little city of Port Deposit, where the Susquehanna empties into Chesapeake Bay, Mayor Hubert Ryan knew how to deal with floods. The community, perched at the river's edge beneath towering cliffs, often experienced ice jams and spring freshets that backed the river up into the city's streets. Long-time residents recalled the 1936

St. Patrick's Day flood when basements filled with river water and the city's marina washed away. But even then and ever since, the great dam a few miles upriver protected them from the worst of the Susquehanna.

Conowingo Dam was one of the big employers in Port Deposit, and word spread among families of the workers that some of the spill gates were being opened. The controlled spill was sure to raise the river and flood riverside homes and businesses. Many residents knew to move their furniture upstairs and pack their cars for a quick getaway, and Mayor Ryan dispatched his small police force to alert the rest. Buses were contracted to take residents to safety at the Navy Training Center at nearby Bainbridge. At midnight Friday, Ryan and his family were the last to leave.

* * *

Dam superintendent English received the RFC's new forecast at 3 p.m., again relayed from Civil Defense headquarters in Harrisburg. He scanned down the blocky print of the teletype: O.D. White predicted a flow of 1,100,000 cfs the next day. This was 230,000 cfs greater than the 1936 flood, and 300,000 cfs more than the dam was designed to handle.[2]

Out on the headworks, Dave Goodland no sooner opened his tenth spill gate when his walkie-talkie squawked anew. "Keep going, Dave. The supe says open them all!" Astonished, Goodland and a fellow dam operator guided the big crane to the next gate. They worked sixteen hours until all fifty-three gates were opened.[3]

* * *

Maryland governor Marvin Mandel arrived at Conowingo Dam to meet with superintendent English and local officials late Friday afternoon. The governor, shaken after touring the destruction around Ellicott City, knew that a dozen people died in the rampaging Patapsco and Patuxent Rivers. Now, Mandel and English huddled at the south end of the dam, facing what could be an even greater disaster. Plumes of spray geysered more than a hundred feet high as seven million gallons of water per second spilled through the gates and crashed onto the rocks below. English gestured toward the far end of the dam and shouted over the roar. "I have men out there now, opening the last of the gates. Governor, it might not be enough. We're near the point where the stability of the dam can't be controlled."[4] The violent shaking of the ground from the relentless pounding of water on the rocks below the dam hammered English's point home.

Mandel's stomach went hollow. He'd already seen enough destruction. "Can't be controlled? Is the dam going to hold?"

English chose his words carefully. "If it gets to 111 feet[5] ... and we expect it will in the next few hours ... we just don't know if the structure of the dam may give."[6]

Mandel declared an "impending disaster" for the entire area below the dam. If failure of the dam seemed imminent, everyone in Port Deposit, Havre de Grace and Perryville would have to leave. Evacuation of residents who lived nearest the river began immediately, and the National Guard moved twenty helicopters to nearby Edgewood Arsenal, on standby to lift anyone remaining to safety.[7]

Conowingo is a "masonry gravity" dam, held in place by the massive weight of its concrete. It was not designed to take the kind of pressure the river was putting on it this Friday night. The dam started to lean – just inches, but leaning – and threatened to break free from its bedrock foundation. The entire structure shook "like a vibrating treadmill".[8]

Walking the headworks, dam operators Dave Goodland and John Ulrich watched as trees, shattered homes, carcasses of cattle, even cars plummeted through the gates to be pulverized in the maelstrom below. Goodland and Ulrich were soon joined by the dam's supervising engineer. With a can of spray paint the engineer marked two lines of little circles in the road – U.S. 1 – atop the north end of the dam. "Get the drill down here. I need you to put two-inch holes twenty feet deep at my marks."

"Sir? Drill holes in the dam?"

"Right. Don't spread this around. If the dam isn't going to hold we've got to make a new spill gate. With dynamite."[9]

The dam operators towed the trailer-mounted drill from the maintenance yard to the road atop the dam. The air was thick with spray that spouted up from a hundred feet below. Vibration from the thundering spill and the stench of the foul water made them nauseous. Goodland and Ulrich maneuvered the drill to the next mark, and the next, until a few hours later they'd drilled the twenty holes. A fifty-foot stretch of the north end of Conowingo Dam was ready for explosives.

If the dam were blown open the sudden gush of flood water through the breach would erase Port Deposit and Havre de Grace from the landscape, sacrificed to save the dam itself.[10] Only the governor could make such a grave decision, so Mandel stayed the night in nearby Edgewood to await word from English. If the flood neared the crest the dam, Mandel would order that the north end of the dam be blown open. Rumors about the dynamite began to quietly circulate in the down-river towns and in the press. "I never heard one way or another whether it was true or not," recalled one reporter.[11] The secret was mostly kept until the full story could be told years later.[12]

Through the long night Superintendent English watched the meter in the control room. Even with all fifty-three spill gates wide open the reservoir kept rising: 110, 110.5, 111 feet – the height at which English feared for the integrity of the dam – were passed. Finally, Saturday morning, the reservoir crested at 111.5 feet, a couple of feet below the top of the dam.

English put in a call to Governor Mandel. "I think we can stand down. She's topped out and starting to drop." Mandel allowed himself to relax. The dam would not be blown open. Although flood waters ran fifteen feet deep through the streets and into the second stories of many riverside homes, Port Deposit and Havre de Grace were spared extinction.

Discharge of the Susquehanna through the penstocks and spill gates at Conowingo reached 1,280,000 cfs, or 7.5 million gallons per second: 400,000 cfs greater than the 1936 record and far beyond the design capacity of the dam.[13] This was more than double the average flow of the Mississippi River. All fifty-three gates remained opened until Sunday afternoon, more than forty-eight hours, when the river finally began to recede.

LAST GASP

Sunday, June 25, 1972
Maniwaki, Quebec

The Ottawa regional weather forecast called for "more rain from Agnes"[1] as the remnant "tropical disturbance" drifted north into Canada. Three to four inches of rain had already fallen in Ontario and Quebec, with another inch or two predicted. It was the most June rain in thirty years.[2] Although local streams ran high, Agnes would not bring the disastrous floods it had to the States.

In Maniwaki, seventy miles north of Ottawa, Allan Reavey picked up supplies for his family's cabin. His wife and son didn't want to endure the two-hour trip from camp, so Reavey made the rainy drive alone. The wind increased as Reavey stepped from the door of his modular home with a heavy box of groceries in his arms. He teetered in the gust before he reached his car, then heard a roar of wind and the crackle of shattering wood behind him. Reavey turned to see his house splinter into a million pieces.[3] A freak tornado had descended on Maniwaki. "It was unbelievable. The whole house went up in the air and dropped on the road like a box of kindling."[4]

Across the street, six-year-old Maria Gough looked from the window and called to her mother. "Mama, the flames and wind and paper are coming!"[5] What was coming was the mobile home of their neighbor Jeanine Lacroix, flying through the air from 250 feet away. Sparking power lines dragged with it as it sailed across the street. There was no time to react before the Lacroix home smashed full force into the end of Gough's. Little Maria and her mother were not hurt, but Jeannine Lacroix died instantly when she touched the steel wall of her shattered home, electrified by the live wires.[6] Lacroix's seventeen-year old son Roger also died, and two younger children seriously injured.

The Maniwaki tornado was a freak phenomenon, spawned in the "death throes of tropical storm Agnes."[7] Canada's weather service didn't see it coming. "We couldn't forecast it if it happened again," a bewildered forecaster at the Uplands Weather Center recalled.[8] There were no obvious thunderclouds in the vicinity, and the storm that spawned the tornado wasn't seen on radar.[9] The short-lived funnel looked "like a pipe in the distance".[10] It left a path of destruction only a few hundred feet long,[11] but in that path ripped twenty homes and businesses apart, killed two residents, and injured eleven more. Shingles, siding, debris and belongings were scattered more than a mile from their origins.[12] Throughout Ontario and Quebec winds from the tail end of Agnes toppled trees and lashed at power lines, and as many as 300,000 homes were without electricity.[13]

* * *

Sunday, June 25, 1972
Washington, DC

Rock Creek returned to its banks Sunday morning, but still ran high and brown with silt. The flood obliterated most of the picnic areas and playgrounds along Beach Drive, and the narrow flood plain along the creek was plastered with slick mud. Still, a few residents ventured along what was left of the creek-side trails, some to gawk at the damage, others for a reprieve in the leafy forest.

Rudy Juarez taught school in Guatemala before coming to Washington, where he worked for better wages as a waiter in an upscale restaurant. Juarez was happy in Washington, and played as the all-star goalie on the Guatemalan community's soccer team. On this sunny afternoon Juarez took his wife Maria and two young boys for their customary Sunday stroll through the forests of Rock Creek Park, unaware of the beating Beach Drive and the park trails suffered in the flood. They slogged through the thick ooze that obscured the once-familiar pathways. The boys complained about the mud getting into their shoes, so Juarez and his wife hoisted them up to carry piggy-back. When they reached what they called their "picnic rock" at the edge of the creek, they ventured onto it for a better look at the rushing brown water.

The big flat rock tilted toward the creek, and its muddy wet surface was slick as grease. Juarez lost his footing, slipped into the water, and dragged Maria with him. As they fell, the boys, Rudy Jr., age four, and Alejandro, age two, let go of their parents. Juarez and his wife scrambled out and raced down the riverbank, calling the boys' names, but there was no sign of them. When they reached a ruined picnic area Juarez found a pay phone – somehow it worked – and called the police. Juarez spoke fluent English, but in his panic shouted frantically in Spanish. The mis-communication delayed rescuers, although it wouldn't have made a difference. Searchers found the body of Rudy, Jr. tangled in brush a quarter mile downstream, but never recovered little Alejandro.[14] Their deaths were the last directly attributed to Hurricane Agnes.[15]

* * *

July 2, 1972
Aboard *Gutenfels*, the North Atlantic

Five days out of Philadelphia with a load of specialty steel, Captain Hans Zimmermann directed his helmsman to turn his 512-foot freighter into the heaving seas. Waves twenty-five feet high, propelled by gale winds, made the *Gutenfels* roll abeam and threatened to shift the cargo in the hold. Zimmerman had seen much worse in his twenty years as captain, and if he didn't take chances his ship would have no problems. But time was money. The turn added hours to his passage, and the owners might grouse about his late arrival at Bremerhaven. "They'll get over it", Zimmerman chuckled. The safety of his ship, his crew, and his cargo were his paramount concern.

Four days earlier the remnant of Hurricane Agnes swirled into the Atlantic near Cape Breton, Nova Scotia. Yet again the storm deepened, kicking up fifty-mph gales as it zig-zagged across the open ocean. Between July 1 and 3 it looped between Northern Ireland and Iceland, disrupting the voyage of *Gutenfels* and a half dozen other ships.[16]

At last, on July 6, the cyclone that began as a cluster of thunderstorms off the coast of Ecuador a month earlier disintegrated in westerly winds near the Faeroe Islands.[17]

Agnes was no more.

AFTERMATH

Sunday, June 25, 1972
Wilkes-Barre, Pennsylvania

On Saturday afternoon the Susquehanna River at Wilkes-Barre crested at 40.9 feet, just an inch lower than Gwinner's emergency Friday morning prediction. The river poured over the flood walls for more than forty-eight hours, even as it slowly receded, until noon Sunday.

Nick Souchik stayed on duty from Thursday to Sunday, grabbing only catnaps on a cot at Civil Defense's ad hoc headquarters in the school administration building. Flood water nine feet deep inundated the roads to his home in Plains, and rose to the ceiling in his parlor, living room, and kitchen. His wife Olga safely evacuated to stay with her sister.

At 5:00 p.m. Sunday, Souchik returned to the courthouse, driven by a National Guardsman in a jeep. He had to see whether his offices in the basement survived the flood, and wanted to check his river gauge. The route to the courthouse zig-zagged through streets that now seemed like a battleground. Huge slabs of broken asphalt and concrete lay in piles where the turbulence of the flood scoured them up from streets and sidewalks. Water gurgled through gullies cut deep into the soil along and across what days before had been busy thoroughfares. Utility poles and traffic signs tilted at wild angles or lay supine, tangled with a hundred miles of electric and phone lines. Rippled drifts of mud and sand piled four feet high at intersections, and spilled into the doorways of homes and businesses.

A line of mud fifteen feet high on every structure marked the crest of the flood. Windows and doors had imploded, allowing flood waters to swirl inside. The front porches of homes in the river's path collapsed, some to remain in a heap, many leaving only a scar where the porch tore away. Some homes cleaved in two, their insides exposed to open air with furniture and belongings still inside, appearing like giant doll houses. Others lifted from their foundations or had been battered into rubble. Muddied cars floated onto front yards where they piled up like toys.

What Souchik did not see was people. A hundred thousand souls fled to high ground, to the relief centers, or to friends and relatives on the hill. There was no sound but for the gurgle of draining water. Not even birds. It was eerie.

A sea of brown-grey muck surrounded the courthouse. Broken trees, the shattered roof of a barn, an overturned boat, thousands of sodden sandbags, and hundreds of dying fish littered the grounds. Incredibly, the sandbags around the foundation and doors of the courthouse, piled at the last minute by National Guard and volunteers, stood proud. A few feet of water seeped into the sub-basement, but the venerable old building didn't suffer much damage.

Souchik slogged through the ankle-deep ooze to his gauge house near the flood wall. He had not measured the river since it spilled over the wall Friday morning. A ring of mud higher than the door ran around the little shed, marking the height of the flood. Inside, the electronic bubble gauge was infiltrated with mud and useless, but the tape measure gauge worked fine: 32.2 feet, now four feet below the top of the wall. He would report it to the RFC when he got back to headquarters.

After reading the gauge Souchik peered over the battered steel flood wall. Beyond it the river raced by, still brown with silt, still carrying a load of debris. There were only concrete piers and a tangle of steel where the North Street Bridge stood just days before. He knew the bridge had collapsed Saturday, but it shocked Souchik to see the wreckage.

Even as Souchik and his escort navigated the soggy, torn streets away from the courthouse, National Guard troops and Red Cross volunteers arrived from their regional depots to spread into the devastated community with supplies, equipment and gritty willpower. The cleanup and recovery began.[1]

* * *

*Streets were scoured away and homes smashed in every community
hit by Agnes, as here along Wilkes-Barre's Charles Street* (U.S. Army
National Guard)

The same scene awaited residents in all the cities hit by the flood:
Richmond, Reading, Corning, Elmira, Wilkes-Barre, Harrisburg,
Lancaster, York, Wheeling, hundreds of smaller communities, even
the big metropolises of Washington, Philadelphia, Baltimore, and
Pittsburgh. Never had a flood devastated such a huge area and so
many people. President Nixon declared all of Pennsylvania, and
most of Florida, Virginia, Maryland and New York disaster areas.

The victims and evacuees had to be sheltered and fed. The
shattered homes and businesses, the mountains of debris, and
the ubiquitous slimy mud had to be cleared away. Electric power,
phones, gas, water and sewers had to be restored, streets and bridges
reconstructed. Only then would it be safe for people to return, repair,
and rebuild, if they could at all. Life had to start over. It would take
many months, even years.

* * *

Most urgently, the flood over-ran public water and sewer works. Electricity for private wells was out, so even areas not flooded lacked a water supply. The National Guard's first order of business was to truck in tankers of potable water. Residents lined up with milk jugs, buckets, and coolers to tote back home. There wasn't enough, and fights broke out among frazzled residents over the meager rations.

Beverage distributors sent bottled water long before Evian and Poland Springs became popular. National Guardsmen at Wilkes-Barre, hot and sweaty from cleanup duty, grinned as a truck from Anheuser Busch pull into their depot. Their smiles grew wider when they opened the door to see sixty pallets marked "Budweiser." This was exactly what they needed! Then, when they opened the first box off the truck their smiles evaporated – the bottles held only water. "Tough luck, men," said the lieutenant in charge, "Send this to the evac center at the high school. They need it more than we do."[2]

Eight million donuts, thirty million cups of coffee and other drinks, and four million meals were served up to the refugees in Wilkes-Barre alone.[3] Throughout the Middle-Atlantic states the Red

Potable water was hard to come by in the entire region hit by Agnes (U.S. Army Corps of Engineers, Tropical Storm Agnes!)

Cross, Salvation Army, local churches, civic groups, and thousands of volunteers staffed the evacuation centers, working tirelessly to feed and shelter the hungry and homeless. Amish and Mennonites came from their farms and shops to serve up their special brand of charitable hospitality at the evacuation centers; others helped haul debris, using mules and sledges where modern equipment couldn't reach.

* * *

The tally of fatalities attributable to the Agnes storm and floods, indeed any major disaster, is a fraught endeavor. Contemporary reports may be based on unfounded rumors, mis-identification, double-counts, or otherwise proved wrong, while later reviews can omit or overlook specific events. Whether a death was "caused" by the flood is a judgment call; some people died while cleaning up after the waters receded, and at least six persons – pilots and reporters covering the storm – died in helicopter crashes. Perhaps the best analysis was performed by Professor Wayne Blanchard years after the floods passed in his review of "Deadliest American Disasters and Large Loss-of-Life Events".[4] According to Blanchard, Agnes claimed the lives of 147 people in ten states, plus at least sixteen in Cuba and two in Canada – every one of them a lost husband, wife, father, mother, son or daughter.

* * *

"Be prepared for a shock," General Townend warned residents before allowing them back into their neighborhoods. Tens of thousands of people lost everything, material and sentimental possessions, often their very homes. "But you're alive, and that's the main point."[5]

Even if a home or business remained intact, the furniture, appliances, carpet, and clothing could not be salvaged. Particularly painful for families was loss of personal papers, pictures and heirlooms. The very plaster and wooden frames of the walls,

Residents and businesses put their destroyed furniture, merchandise, and debris at curbside during the day; city and National Guard trucks hauled it away at night (U.S. Army Corps of Engineers, The Corps Responds)

infected with mold and toxins, had to be ripped out. Residents and shopkeepers piled their trash and debris by the street during daylight; city and National Guard trucks trundled by at night to haul it away. In Wilkes-Barre alone *eight million cubic yards* of debris from ruined homes and businesses was swept up and dumped at the landfill.[6]

Throughout the Middle Atlantic region dashes of color stood out against the monochrome backdrop of mud and devastation. Amidst the muck and sorrow, people flew their American flags, bright banners waving from windowsills, porches, trees and clotheslines. Although the people might be down, they must not be counted out!

Everywhere the rivers went, *everywhere*, a slimy, toxic mud remained behind. Not just ordinary river silt, the mud was contaminated with sewage from swamped sewer plants, manure and fertilizers from agricultural lands, gasoline from ruptured tanks, chemicals from over-run refineries, rotting plants, dead fish. In the coming days residents added spoiled food and moldy plaster to the dreck. Scavenger worms appeared in the ooze, writhing underfoot as volunteers skidded through it with shovels and brooms. Everyone called it "flood mud," filling streets and homes, emitting a fetid stench no one would ever forget. "And then there was that smell," Wilkes-Barre councilman Tony Brooks said years later. He had been only eight years old at the time of the flood. "It was so awful, regardless of your age you can remember the smell of flood mud."[7]

The mud dried and turned to dust as the waters ebbed and the sun came out. Wind and passing vehicles kicked it up into toxic clouds that penetrated clothing, eyes, and nostrils. Public Works ironically had to spray the streets with water keep the dust down.

At Elmira, college student Irene Fellis was home for the summer. When her family heard the evacuation order they went to stay with her grandfather up the hill. Four days later they returned to their home to find devastation. "Our house looked like the Valley of Death. Everything, inside and out, was overturned, broken and destroyed. The water went halfway up the stairs, and there was mud, mud, mud everywhere. And, oh that smell!" They received Red Cross vouchers for daily needs and a subsidized loan, but the Fellis family didn't know where to begin. "Mom just sat under a tree in our yard, crying and praying."[8]

Then – as they would hundreds of thousands of times – the angels appeared. Six couples from a church at Syracuse came to the the Fellis neighborhood every weekend, all summer, to help. They brought shovels, mops, rakes, buckets, wheelbarrows, and bags overstuffed with clothes. "My mother told them she had no money to pay them, but they said they didn't want money, they only wanted to help us. I had never experienced such kindness from complete strangers, and I will always remember them."[9]

* * *

At the front lines of flood response everywhere stood the young people. "Thank God for the hippies," said an older volunteer.[10] In Corning, a "Youth Emergency Service" sponsored by Corning Glass,

Ingersoll Rand and the school district cleaned up and repaired broken houses and major appliances. Over the summer, four hundred "YES kids" rehabilitated more than 4,500 homes. Their gusto changed the way residents looked at the young people. One middle-aged homeowner watched as a bearded, blue-jeaned teenager shoveled mud out of his living room. "If I hear anyone say anything bad about these hippies again, I'll punch him in the nose!"[11]

At the Wilkes-Barre/Scranton Airport, a distribution center for flood relief, Scranton Mayor Eugene Peters watched a thousand long-haired volunteers unloading supplies from military flights and delivering food and clothing to flood victims. "Last week I denounced them as hippies. But today I see them as the greatest young people in the world."[12] "Teeny-boppers toiled, then collapsed in each other arms," said response coordinator Rep. Daniel Flood. "Nobody can say anything bad to me about those little bastards ever again!"[13]

* * *

The flood left tens of thousands of residents homeless, their houses swept away or shattered in the raging waters. Expanding a program begun after Hurricane Camille three years before,[14] the Department of Housing and Urban Development trucked mobile homes, more than 11,000 of them, to the stricken communities as temporary shelters in Richmond, Wilkes-Barre, Corning. Everyone called them "HUD huts," years later to be known as "FEMA trailers." Some were placed on residents' lots after the debris was cleared away, some were set up in camps outside of town. The trailers were not luxurious, but they were free, and for many displaced flood victims were home for more than a year.[15] It was the first mass deployment of mobile homes in a natural disaster, a program that continues to this day.

* * *

At first there was little organization. Police, firemen, and volunteers spontaneously saw what had to be done, but it was *ad hoc* with no one in charge. Not everything went smoothly. The overwhelming needs spanning such a huge area stressed government, volunteers and civic agencies to the breaking point. In many places there was chaos. Weary victims stood in long lines, some with bawling toddlers in tow, to get help. Everything took longer, much longer, than residents and local officials thought it should. Sometimes relief agencies competed for supplies, equipment, and control. Judge Max Rosen, chair of a flood recovery task force at Wilkes-Barre, put it this way in testimony to Congress:[16]

> *Time does not permit me to describe the Herculean and magnificent struggle, nor the chaos, confusion and anger that ensued after the flood waters receded, because there was no single agency with authority to plan or lead the recovery effort.*

There was, especially, no money. A 1968 law made flood insurance available, but of the 20,000 homes damaged in Luzerne County only two were insured. Many flood victims were at the end of their rope. One victim could have been speaking for hundreds of thousands others when he wrote to President Nixon. "My mind is full of problems, my heart is full of disgust and disappointment. I've lost everything. I am broke, disquieted, heart-broken, and on the verge of mental and physical collapse."[17]

Leaders emerged as the days turned into weeks and months. Corning Glass chairman Amory Houghton put the resources of his company behind the cleanup in Steuben County, and personally took command. Congressman Dan Flood – "it takes a Flood to beat a flood" – championed a national response to the disaster with the Agnes Recovery Act, rushed through Congress only two months after the storm. The new law provided short term loans, desperately needed, to flood victims. President Nixon, who visited the flood-stricken region three times, said the law would bring the crippled cities back "better than ever."

Elliot Knauer, Pennsylvania's young deputy Secretary of Public Welfare, set up "one-stop shopping" relief stations in the flood zones where victims could access every kind of service they needed. Entertainer Bob Hope hosted a national telethon to raise money for the Red Cross. President Nixon appointed Frank Carlucci – one of his top aides and a son of the Wyoming Valley – as "flood czar" to coordinate and expedite federal aid. Many of these initiatives became standard practice in disasters to come.

* * *

Local volunteers, art students from across the country, and experts from around the world descended on Corning's Museum of Glass to salvage and restore the priceless collections. Piece by meticulous piece they cleaned and re-assembled fragile glass exhibits from the mounds of shattered shards. Science director Robert Brill rushed in refrigerated trucks to freeze the ancient books, archives and photographs to stop the growth of mold. Page by page the volunteers dried, pressed, and preserved precious documents. The museum triumphantly re-opened only a month after the floods, though it would be years before the collections, to the extent they could be, were restored.[18] The recovery of the museum and its collections became the model world-wide for salvage of damaged glass and restoration of water-logged documents.

Eventually the Trustees rebuilt the Museum of Glass on concrete stilts, an architectural gem with the most valuable collections housed high above flood level. A sign halfway up the panoramic window in the lobby, overlooking the gift shop and food court twenty feet below, prominently demarks "High Water – June 23, 1972".

* * *

Tommy Hilfiger's little People's Place was the only shop on Elmira's Water Street to survive the flood. "While all these clothing stores, boutiques, gift shops and restaurants were wiped out, we

were the only store in town with dry merchandise."[19] Residents
of Elmira who lost everything needed new clothes, and flocked to
People's Place as soon as the streets cleared. Hilfiger sold his entire
stock within hours, then went to New York City's garment district
to buy a truckload of new merchandise. He sold it all at a fair price,
often paid with Red Cross vouchers. Tommy Hilfiger, Inc. was on
its way to becoming a world-wide giant of design, fashion and
entertainment.

* * *

Officials at Panama City, Florida, wanted to sue the Weather
Service and news media for "crying wolf" about the predicted
impact of the storm. Thousands of tourists fled the beach resorts
in the face of what Rep. Bob Sikes called "highly exaggerated"
warnings.[20] Some headlines portended 120-mph winds at Panama
City, but Agnes brought breezes of only forty-five mph and caused
little harm. The mayor estimated his city suffered $100 million in
lost tourist dollars on account of the fizzled forecast.[21] Residents
and tourists "will soon start disbelieving what they hear from the
Weather Service, and refuse to evacuate when told to," complained
Sikes. "Then a major storm will come along and many people will
die."[22]

In fact, National Hurricane Center Director Bob Simpson's
forecast for landfall was accurate: tropical storm winds and a
moderate storm surge near Apalachicola, well east of Panama City.
Apalachicola and nearby coastal villages indeed suffered serious
coastal flooding and power outages, and passed a resolution thanking
the Weather Service for its "timely and accurate assessments".[23]

While Panama City complained that warnings about Agnes
were overblown, Congressman John Heinz at Pittsburgh beefed
that the forecasts were too little, too late.[24] Heinz was incensed that
the Ohio RFC closed up shop for the night, saying that "if they
stayed on duty, a lot of damage could have been avoided."[25] He
convened a hearing to grill Weather Service administrators, where
NOAA director Robert White explained that Agnes was "so erratic

in its motion it was unpredictable".[26] The Ohio RFC, he said, "gave prudent warnings with the information they had at the time".

* * *

Agnes could be considered one of America's most, if not *the* most, influential storm ever,[27] leading to far-reaching changes in the way the country plans, prepares, and responds to natural disasters. Much as flood *protection* became a matter of federal action after 1936's St. Patrick's Day, disaster *response* became a federal matter after Agnes.

Frank Townend's Council on Civil Defense had money for underground "fallout shelters", but nothing for disaster evacuation centers. Too focused on military matters at the expense of natural catastrophes, after Agnes the Council evolved into the Federal Emergency Management Agency[28] – FEMA – to consolidate the response functions of dozens of agencies and coordinate federal and state responses to all manner of disasters. FEMA would become most prominently known to the public as manager of the Hurricane Katrina response in 2005.

Likewise, the federal Agnes Recovery Act gave $5,000 cash to families whose homes had been lost or damaged, plus low-interest loans for rebuilding. In ensuing years Congress expanded the program and made it permanent, extending federal aid to victims of disasters nationwide.

* * *

The railroad industry in the east was suffering even before Agnes came along. Carriage of bulk cargo, especially coal, had declined. Shippers sent goods by truck over the new interstate highways, and passenger rail service gave way to air travel. Agnes dealt the industry a knock-out blow. On Penn Central's system alone 1,400 locomotives, 2,291 freight cars, forty-eight bridges, and four thousand miles of track and grade were damaged or destroyed.[29] Penn Central was already in bankruptcy, and the storm dashed any hope of recovering. The Erie Lackawanna, faced with damage nearly as bad as Penn Central, filed for bankruptcy three days after Agnes passed. The flood was the "coup de grace, and without the money needed for the cleanup, it was the end," lamented Erie chairman John Fishwick.[30] Many smaller lines threw in the towel, never to reopen – some of which are now popular "rail-trails" for hikers, cyclists and equestrians.[31]

Emergency legislation after Agnes created the Consolidated Rail Corporation – "Conrail" – a quasi-public corporation to rescue the flagging lines. For the next twenty-five years, Conrail operated freight and commuter rail service throughout the Northeast, mid-Atlantic and Great Lakes states, until sold to CSX and Norfolk Southern in 1997.

* * *

The dikes along the Allegheny, Genesee, Chemung, Schuylkill, Susquehanna and James Rivers had been designed to protect against what engineers presumed could be the worst possible flood, something higher than even the great St. Patrick's Day disaster of 1936. Agnes exploded their presumptions. The rivers rose far higher than ever before, sometimes more than doubling the prior records. "We never thought the water could come over the walls," was a common refrain throughout the mid-Atlantic region.

If the old walls weren't high enough, then the cities would build them still higher. And so they did, at Corning, Richmond, Sunbury, Wilkes-Barre and other communities. Luzerne County raised its dikes from thirty-seven feet up to forty-three feet, high enough for another Agnes. The walls were extended to sixteen miles, supplemented by massive pumping stations and broad flood retention basins, making Luzerne County's river defenses among the most extensive nationwide.[32] Almost everywhere the new flood works are enhanced by gardens, picnic areas, amphitheaters, boat launches, pedestrian and bike paths. The walls are no longer necessary eyesores.

Other works were built to stave off future floods. Big dams on the Cowanesque and Tioga Rivers, in the headwaters of the Chemung, long contemplated but built with urgency after Agnes, promised to reduce flooding downstream. Wetlands were rehabilitated to better absorb flood waters, and trees planted to inhibit runoff. In some parts of the flood zone – *e.g.* at Elmira – authorities removed structures damaged in the flood, and left the acreage as open space.

So far, the new, higher walls and other works have staved off catastrophe. There hasn't been "another Agnes", but no one says "it could never happen again!"

* * *

The Susquehanna River contributes half of all the water that flows into Chesapeake Bay, the largest estuary in America. Normally, even during floods, the load of silt it carries gets trapped behind Conowingo and smaller dams at Holtwood and Safe Harbor. Agnes broke all the rules of "normal." At the height of the flood, seven-and-a-half million gallons of water spewed through and over Conowingo Dam *every second*, half again as much as ever before, carrying with it an enormous burden of silt, fertilizer from farms and lawns, sewage from municipal utilities, and contaminants from industry. Instead of being trapped behind the dam, the load of silt and pollution plunged into the bay. Worse, in the "mother of all scouring events,"[33] the flood tore twenty million tons of accumulated sediment from the bed of the Conowingo reservoir and added it to the load.

The rush of fresh water from the Susquehanna, with lesser but still enormous contributions from the Potomac, Patuxent, Patapsco, Rappahannock and James Rivers, pushed the bay's salt water boundary far south, killing the creatures that thrived in brackish water. Sunlight couldn't penetrate the murky water. Fertilizers washed from fields upstate forced a bloom, then rapid decomposition, of underwater grasses. Already stressed by pollution and over-fishing, the ecosystem of Chesapeake Bay collapsed.[34] A "dead zone" extended from the mouth of the Susquehanna sixty miles down to St. Michaels, the grasses and most marine life wiped out. The oyster industry, long a mainstay of the Chesapeake economy, was decimated; oysters polluted by sewage pouring down with the flood couldn't be sold, and the remaining population crashed.[35] Ninety percent of the soft-shelled clams died. Maryland's famous blue crabs, nursed in the sanctuary of sea grasses, disappeared. Years went by before the grasses rebounded, decades before marine life regained a foothold. Even fifty years later, the oyster and crabbing industries remain a shadow of what they were before Agnes.[36] St. Michaels waterman Larry Simmons reflected on the disaster. "I didn't see how one storm could change my life. But then I watched my bay turn brown, with dead cows and drums of chemicals floating by, and I knew that my best fishing days were over."[37]

In five days Agnes washed more sediment and pollution down the Susquehanna River and other tributaries of Chesapeake Bay than had come in the previous fifty years.[38] Nothing like it had occurred in the thousands of years the estuary has existed.[39] The sediment settled to the bottom in a distinct black deposit eight inches thick. Over eons it will compress and turn to shale, perhaps to be uplifted by tectonic forces and exposed again by erosion. If there are geologists a hundred million years from now, they might point to this thin layer of dark rock and say, "*This* was Hurricane Agnes."

GOLD MEDAL

October, 1973
Washington, DC

O.D. White stood at the dais in the auditorium at the Department of Commerce building, adjusting his tie as he looked over the audience. He and the other hydrologists of the Middle Atlantic River Forecast Center were there to receive a "gold medal citation," the Department's highest award.[1] It read, in part:

> *They performed extraordinarily under intense stress, aggravated by simultaneous record-breaking floods from southern Virginia to southern New York, despite loss of electric power, computers inoperative, and gauging stations destroyed.*[2]

Ever modest, White began his acceptance remarks with a joke:

> *I dreamed I died and went to heaven, where Saint Peter asked what I had done to deserve passage through the pearly gates.*
> *'Well sir, I forecast the greatest flood in history.'*
> *Saint Peter smiled. 'Very impressive, but not so fast Mr. White. Let me introduce you to Noah.'*

* * *

The Army Corps of Engineers calculated that Agnes dumped 28.1 trillion gallons of water over the eastern United States,[3] more than ever produced by a single storm, causing the most extensive flooding in U.S. history.[4]

White's joke, though, was true. Other storms, and other floods, would come. In 2011, the remnant of Tropical Storm Lee drenched the mid-south and eastern states with twenty-nine trillion gallons of rain, nosing ahead of the Agnes record.[5] Lee took a path remarkably similar to Agnes, combining with a stalled cold front to sweep up the Susquehanna valley. The headlines screamed *Echoes of Agnes* as riverside cities braced for disaster. But Corning, Elmira and other communities had raised their dikes, and were spared the worst. At Wilkes-Barre, discharge of the Susquehanna peaked at 310,000 cfs, 35,000 cfs less than Agnes. However, with the flow confined within the city's improved flood walls, the stage climbed a foot and a half higher. Residents evacuated as a precaution, but Lee did not top Wilkes-Barre's new dikes and did little damage.

In 2017 Hurricane Harvey, enhanced by a warming climate,[6] slammed Houston and the mid-south with thirty-three trillion gallons of rain,[7] taking the lead in the record book. Harvey impacted a much smaller area than Agnes, but rainfall was measured at up to sixty inches over several days. Agnes fell to third on the list of "most rain".

Agnes also caused more damage than any storm that preceded it.[8] The Corps of Engineers put it starkly: *the Agnes storm was the most costly natural disaster, in terms of property damage, in the history of the world.*[9] Estimates varied, but most pegged damages at about $4 billion[10] – $26 billion in 2020 dollars. President Nixon called it *the worst natural disaster in the whole of America's history.*[11]

The "most damage" record was held by Agnes for twenty years, until 1992's Hurricane Andrew devastated south Florida and Louisiana. Hurricane Katrina in 2005, "Superstorm" Sandy in 2012, and several other storms caused even more damage. When costs are normalized to present-day dollars Agnes is still in the top ten, but barely, and probably won't remain there long.

* * *

Despite the passage of time Agnes has not been forgotten. The trauma of the devastation is written in the psyche of riverside communities, and Agnes remains the flood against which all others will forever be compared. For years, whenever there was especially heavy rain, residents called the River Forecast Center with a shaky voice, "How bad will this be?" Some kept a moving company on retainer, ready to help them evacuate on short notice.

One hydrologist recalls addressing a civic meeting in Luzerne County in the early 1990s, soon after he was hired at the RFC. "I casually mentioned that some storm might be on a track similar to Agnes. It was just a hypothetical, but the audience gasped. An older man stood up and said, 'Are you trying to panic people? Agnes nearly wiped us out, and don't *ever* take that lightly!'"

Newspapers in riverside cities still publish special commemorative editions[12] on the decennial anniversaries of Agnes, and there have been several television specials recalling the event.[13] There was even a children's book, *The Flood that Came to Grandma's House*,[14] to quell the anxieties of little ones who suffered through the disaster. The storm inspired artists, musicians and poets to express their trauma through sketches and paintings, songs,[15] and verse. The 1500-page *Post Flood Report* published by the Army Corps of Engineers – not often regarded for its artistic sensibilities – is extensively punctuated by poetry:[16]

> *The river seemed so tranquil that dreary, rainy day.*
> *No one was even wary a flood was on the way.*

A brighter side of the Agnes tragedy can be seen in the way it inspired some people to embrace public service, particularly understanding and preventing flood disasters. Eleven-year-old Kevin Hlywiak watched with his family as the flood rose toward their home. He knew people who died, and helped clean up [and rebuild] damaged homes around Harrisburg. College student Joe Ostrowski ran from devastation of the Forty Fort Cemetery then joined in the months-long recovery in the Wyoming Valley. Both dedicated themselves to flood forecasting, and went on to long and distinguished careers as hydrologists at the River Forecast Center.

River forecasting changed too. The Middle Atlantic RFC moved to new offices at State College, Pennsylvania, where it operates under nationwide protocols of the "Advanced Hydrologic Prediction Service". There are many more stations reporting river stages and precipitation than there were in 1972, most via automated internet and satellite connections. Ever-more-powerful computers digest vastly more data to run the river forecasting models. High-resolution satellite images and Doppler radar – now a staple of TV weather reports – let hydrologists know just how much rain is falling in real time. There remains, however, a vital network of volunteer observers who report their backyard precipitation[17] or serve as backup to automated river gauges. River forecasting still comes down to dedicated volunteers and skilled hydrologists collecting and analyzing data to stay ahead of the next flood.

Most of the principal players in the Agnes drama are gone now. Hydrologist-in-Charge O.D. White retired to his beloved Scotts Hill, Tennessee, where he passed away in 1996. Nick Souchik, river observer and director of Civil Defense at Wilkes-Barre, died in 1984 at age 68, his widow Olga in 2010. Hydrologist Mike Gwinner's 2018 obituary noted his Gold Medal Citation for *saving many lives with exceptional forecasting during Hurricane Agnes*. And General Frank Townend, commander of Civil Defense in Luzerne County, awarded a Purple Heart and Bronze Star for valor in the Battle of the Bulge, recalled the week of June 22-26, 1972, as the most desperate hours of his life.[18] He died in 2001, and is buried in the old Forty Fort Cemetery alongside four generations of his ancestors, not far from the monument remembering the 2,500 graves gouged out by the flood. Near Townend's own grave is a simple marker identified only as "Agnes".

ENDNOTES

PROLOGUE

1. "Half-Million See President," *Times-Leader*, Wilkes-Barre, Pennsylvania, August 15, 1936, 1.

2. Atkins, Herbert E.: *The Wyoming Valley Floods of 1936* (Wilkes-Barre, Pennsylvania: Collins Press, 1936), 9.

3. William H. Shank, *Great Floods of Pennsylvania*. York, Pennsylvania: American Canal and Transportation Center, 1972, 35-46.

4. $4.75 billion in 2020 dollars.

5. Arnold, Joseph L. *The Evolution of the 1936 Flood Control Act* (Fort Belvoir, Virginia: U.S.Army Corps of Engineers, Office of History, 1988), Preface.

6. "Johnstown Granted WPA Fund to Speed Control Dam Plans," *The Johnstown Chronicle*, Johnstown, Pennsylvania, August 14, 1936, 2.

7. "Havoc Left by Flood Heaviest in History," *Times-Leader*, Wilkes-Barre, Pennsylvania, August 15, 1936, 2.

EL NINO

1. Cesar N. Caviedes, "El Nino 1972: Its Climatic, Ecological, Human and Economic Implications," *Geographical Review* 65, no. 4 (October 1975).

2. Jerome Namias, "Birth of Hurricane Agnes – Triggered by the Transequatorial Movement of a Mesoscale System Into a Favorable Large-Scale Environment," *Journal of the American Meteorological Society* 101, no. 2 (February 1973): n. 3.

3. "It's Our Kind of Conspiracy," *New York Times*, March 26, 1972, E-2.

RIVER OBSERVER

1. Accursia, Sister M., "Polish Miners in Luzerne County, Pennsylvania," *Polish American Studies* 3, no. 1-2 (1946).

2. In 1972 most stations measured the river stage with a float under the gauge house that rose and fell with the river's surface. Others used a wire-weight gauge. The digital bubble gauge at Wilkes-Barre was cutting edge technology for the time.

3. *e.g.*, Robert R. Mason, Jr. and Benjamin A. Weiger, *Stream Gaging and Flood Forecasting: A Partnership of the U.S. Geological Survey and the National Weather Service*, U.S. Department of Commerce, National Oceanic and Atmospheric Administration and U.S. Department of the Interior, U.S. Geological Survey, August, 1995, accessed June 20, 2021, https://pubs.usgs.gov/fs/1995/0209/report.pdf.

FLASH FLOOD

1. The death toll in the Rapid City flood ultimately tallied 238. Janet M. Carter, Joyce E. Williamson and Ralph W. Teller, "The 1972 Black Hills-Rapid City Flood Revisited," United States Geological Survey, Fact Sheet FS-037-02 (2007).

2. Ibid.; Also, U.S. Department of Commerce, National Oceanic and Atmospheric Administration. *Black Hills Flood of June 9, 1972*. Natural Disaster Survey Report 72-1 (August, 1972).

3. NOAA, *Black Hills Flood*, 1.

4. Ibid, at 9. "No organized real-time system for reporting river and rainfall data exists in the Black Hills area. ..."

5. James W. Wilson, "Evaluation of Precipitation Measurements with the WSR-57 Weather Radar," *Journal of Applied Meteorology* 3, no. 2 (1964): 164-174.

6. NOAA, *Black Hills Flood*, 7

7. Namias, *Birth of Hurricane Agnes*: "Apparently, the large-scale convective cloud system associated with the cold front, the general cyclonic vorticity field, and the implied high moisture content aloft provided a favorable large-scale environment for triggering the hurricane by the nucleating intense meso-scale cloud system arriving from the south over anomalously warm water."

   Also, Robert H. Simpson, and Paul J. Hebert, *Atlantic Hurricane Season of 1972*, U.S. Department of Commerce, National Oceanographic and Atmospheric Administration, National Weather Service (1973).

SALVANOS!

1. Robert H. Simpson. *Hurricane Pioneer: Memoirs of Bob Simpson* (Boston: American Meteorological Society, 2015).

   Also, Robert Simpson. Interview by Edward Zipster, American Meteorological Society, 1989, accessed Sept. 28, 2021. https://opensky.ucar.edu/islandora/object/archives%3A7642.

2. John Hope became familiar to viewers of TV's The Weather Channel as its hurricane expert after he retired from the Weather Service.

3. "Pinar del Rio and its Inland Geography," *Radio Rebelde*, accessed Sept. 30, 2021, https://www.radiorebelde.cu/english/news/pinar-rio-and-its-inland-geography-20130619/.

4. Depiction a flood of Rio Cuyaguateje in a novel by Juio Antonio del Marmol: *The Havana Conspiracies – Rites of Passage of a Master Spy* (Bloomington, Indiana: Trafford Publishing, 2016).

5. U.S. Department of Commerce, National Oceanic and Atmospheric Administration, National Weather Service, *Preliminary Reports on Hurricanes and Tropical Storms: Hurricane Agnes* (September 1972): 3.

6. "Agnes Kills Seven in Cuba," *The Miami Herald*, June 19, 1972, A-2.

7. "Principales Eventos Pluviales Sobre Cuba en el Periodico 1963-2006," *CubAgua*, (Havana, Cuba: Instituto Nacional de Recurso Hidraulicos, July 23, 2011).

HURRICANE HUNTERS

1. The 53rd Weather Reconnaissance Squadron is now (2020) based at Keesler Air Force Base at Biloxi, Mississippi.

2. Robert H. Simpson and Paul J. Heber, *Atlantic Hurricane Season of 1972*, U.S. Department of Commerce, National Oceanic and Atmospheric Administration, National Weather Service, National Hurricane Center, Miami, Florida, 1973..

3. Ibid.

4. National Hurricane Center Director Simpson observed: "One reason for the lack of storm activity may have been the below-normal sea-surface temperatures in the region from the Lesser Antilles to Africa. ... During this period the sea-surface temperatures in the extreme eastern Pacific were well above normal and tropical cyclones occurred there in greater numbers, larger sizes, and greater strength than normal." Simpson, *Atlantic Hurricane Season of 1972*.

5. "Renewed Flood Threat Eased, One Killed," *Rapid City Journal*, June 18, 1972; "Heavy Rains Drench Rapid City Again," *Associated Press*, June 18, 1972.

GENERATION GAP

1. Jack Jarecki, "How the City Wasn't Saved," *Wyoming Observer*, Wyoming, Pennsylvania, July 2-8, 1972.
2. "Hooliganism on the Increase Here," *Times-Leader*, Wilkes-Barre, Pennsylvania, June 19, 1972, 18.
3. Anthony J. Mussari, *Appointment With Disaster: The Swelling of the Flood*. Wilkes-Barre, Pennsylvania: Northeast Publishers, 1974, 48-49, 51.
4. Knauer's remarks at the Sterling Hotel were reported to the author by John Comey, Public Information Officer to Pennsylvania's Council on Civil Defense in 1972.

THE DIRTY QUARTER

1. "Agnes Soaks Cuba; Keys Put on Watch," *Miami Herald*, Miami, Florida, June 18, 1972, A1-A2.
2. "Agnes Roars up Gulf," *St. Petersburg Times*, June 19, 1972, A2.
3. Ibid.
4. "Everything Exploded as Twisters Hit Keys," *Miami Herald*, Miami, Florida, June 19, 1972, 1A.
5. Ibid., 28A.
6. Ibid., 28A
7. John Nelander, "Anniversary: Hurricane Agnes battered Keys, Slammed Panhandle 46 year ago," *Florida Weather Watch*, Palm Beach, Florida, 2018, accessed Sept. 28, 2021, https://floridaweatherwatchblog.wordpress.com/2018/06/19/anniversary-hurricane-agnes-battered-keys-slammed-panhandle-46-years-ago/.
8. Ibid.
9. "Agnes Roars Up Gulf," *St. Petersburg Times*, June 19, 1972, A2.
10. Bartlett C. Hagemeyer, and Scott M. Spratt, "Thirty Years After Hurricane Agnes - The Forgotten Florida Tornado Disaster," Presented at the American Meteorology Society's 25th Conference on Hurricanes and Tropical Meteorology, San Diego, California, 2002, accessed September 30, 2012, file:///C:/Users/grlet/AppData/Local/Temp/34949.pdf
11. Ibid.
12. "Angry Agnes Leaves Trail of Death," *News Tribune*, Fort Pierce, Florida, June 19, 1972, 1.
13. Hagemeyer, "Forgotten Florida Tornado Disaster."
14. "Agnes Careens into Panhandle Area", *Naples Daily News*, Naples, Florida, June 19, 1972, A-2.
15. "Biggest Blow: $4 Million; Two Twisters Strike, Leave 100 Homeless," *Today – Florida's Space Age Newspaper*, Brevard County, Florida, June 20, 1972.
16. Hagemeyer, "Forgotten Florida Tornado Disaster."
17. "Weather," *The Washington Post*, June 18, 1972, C-4.
18. "Five Held in Plot to Bug Democrats' Office," *The Washington Post*, June 18, 1972, 1A.
19. *e.g.*, "Heavy Rains, Winds Lash Area, Hundreds Are Evacuated," *The Washington Post*, June 22, 1972, 1; "East Coast Flood Ebbs," *The Washington Post*, June 25, 1972, A1, A20.

LANDFALL

1. Simpson and Hebert, *Atlantic Hurricane Season of 1972*.
2. Ibid.
3. National Weather Service, *Hurricane Agnes, the 45th Anniversary*. State College, Pennsylvania, 2007, accessed June 20, 2020, https://www.weather.gov/ctp/Agnes.
4. "Weather Bureau Probe is Asked," *Panama City News-Herald*, Panama City, Florida, June 21, 1972, 9.
5. "Hurricane Strikes Cuba, Heads Toward Gulf Coast," *Miami Herald*, Miami, Florida, June 19, 1972, 1.
6. "Agnes Spreads Terror, Calms Down in Georgia", *The Patriot News*, Harrisburg, Pennsylvania, June 20, 1972, 1.
7. Simpson and Hebert, *Atlantic Hurricane Season of 1972*.
8. Ibid. "Less well forecast was the re-intensification of the storm center while over land, the increased intensity of rainfall, and he attendant record-breaking floods."

FLASHBACK: CAMILLE

1. Robert H. Simpson and Arnold Sugg, *The Atlantic Hurricane Season of 1969*. National Hurricane Center, Environmental Science Services Administration (April, 1970).
2. Ibid.
3. Earnest Zebrowski and Judith A. Howard, *Category 5 – The Story of Hurricane Camille* (Ann Arbor, Michigan: Univ. of Michigan Press, 2005).
4. "Unprecedented Rain: Hurricane Camille's deadly flood in the Blue Ridge Mountains," Washington Post, August 19, 2013; Also, U.S. Department of Commerce, National Oceanographic and Atmospheric Administraion, Environmental Science Services Administration, *The Virginia Floods, August 19-22, 1969: A Report to the Administrator* (September, 1969).
5. There is an outstanding narrative of the Camille floods in Virginia at Earl Swift, "Hurricane Camille, Parts 1-5," *The Virginian-Pilot*, Norfolk, Virginia, Aug 19, 2009. Also, Stefan Bechtel, *Roar of the Heavens: Surviving Hurricane Camille* (New York: Citadel Press, 2006).
6. An all-time 12-hour record rainfall of 34.30 inches occurred at Smethport, Pennsylvania, in 1942, 28.5 inches of it in just three hours. Joe Bellini and Bill Kappel, "Flood Analysis for the World-Record-Setting July 1942 'Smethport' Storm," *Journal of Dam Safety* 16, no. 3 (2019).
7. NOAA, *The Virginia Floods*, 2: "The 27-28 inch rainfall within about eight hours in Nelson County represents one of the all-time meteorological anomalies in the United States."
8. "Unprecedented Rain: Hurricane Camille's deadly flood in the Blue Ridge Mountains," *Washington Post*, August 19, 2013.
9. Judith A. Howard and Ernest Zebrowski. *Category 5: The Story of Camille, Lessons Unlearned from America's Most Violent Hurricane* (Ann Arbor: University of Michigan Press, 2007).
10. NOAA, *The Virginia Floods*, 4.
11. Ibid., 20-24.
12. Ibid., 16.

13. "Virginia Floods Kill 38; James River Threatens Richmond," *Washington Post*, August 21, 1969.

14. "Toll Reaches 60 as Floods Recede," *Washington Post*, August 23, 1969.

15. A crest of 30 feet was reached at the Richmond City Locks in 1771 – surely among the oldest official flood records in the United States. NOAA, The Virginia Floods,17; Also, "James BuildsToward 31-Foot Crest: Goodland Declares Disaster," *Richmond Times-Dispatch*, August 22, 1969, 1.

16. "Toll Reaches 60 as Floods Recede," *Washington Post*, August 23, 1969.

AGNES REBORN

1. "Tropical Depression Brings Needed Rain to Crops," *The Robesonian*, Lumberton, NC, June 22, 1972, 1.

2. NOAA: *Preliminary Reports*, 4: "Rains were mostly beneficial".

3. James E. Hudgins, "Tropical cyclones affecting North Carolina since 1586: An Historical Perspective," U.S. Department of Commerce, National Oceanic and Atmospheric Administration, National Weather Service, *National Weather Service Technical Memo* (April 2000); NOAA, *Preliminary Reports 1972*, 4-5.

4. Wayne Blanchard, *Deadliest Natural Disasters and Large Loss-of-Live Events*, August, 2017, accessed June 30, 2021, http://www.usdeadlyevents.com/1972-June-18-26-hurricanetropical-storm-agnes-flooding-east-coast-esp-pa-138-140/; NOAA, *Preliminary Reports 1972*, 10.

5. "Mudslide in Western N.C. Wrecks House," *The Robesonian*, Lumberton, North Carolina, June 22, 1972, 2.

6. "The remnants of Agnes, now over eastern North Carolina, reignited back to tropical storm strength in a relatively rare interaction with a vigorous weather system that had settled in from the Great Lakes region the day before." National Weather Service. *Hurricane Agnes - The 45th Anniversary*, State College, PA, 2017, accessed September 30, 2021, https://www.weather.gov/ctp/Agnes.

7. Simpson and Hebert, *Atlantic Hurricane Season of 1972.*

8. "Agnes Rages to New England," *Washington Post*, June 23, 1972, A14, citing NWS meteorologist David Greenberg: "I can't remember it ever happening. Certainly, it isn't something you would forecast."

9. United States Department of Commerce, National Oceanic and Atmospheric Administration (NOAA). *Natural Disaster Survey Report 73-1: Final Report of the Disaster Survey Team on the Events of Agnes*, Rockville, Maryland (February, 1973), 5.

10. Richard M. DeAngelis, "Hurricane Agnes June 14-23, 1972," *Climatological Data, National Summary* 23, no. 6, U.S. Department of Commerce, National Oceanic and Atmospheric Administration, Environmental Data Service (1972): 313.

11. NOAA, *Final Report*, 5-7.

12. Richard M. DeAngelis and William T. Hodge, *"Preliminary Climatic Data Report Hurricane Agnes June 14-23, 1972,"* NOAA Technical Memorandum EDS NCC-1, U.S. Department of Commerce, National Oceanic and Atmospheric Administration, Environmental Data Service, August (1972), 9.

13. Blanchard, *Deadliest Natural Disasters*, n. 171-173

14. U.S. Department of Commerce, Environmental Science Services Administration. *The Virginia Floods, August 19-22, 1969: A Report to the Administrator*, September (1969): 16-17.

15. U.S. Department of Commerce, *Virginia Floods*, 24.

16. NOAA, *Final Report*, 7-8.

DO YOU REMEMBER?

1. Scottsville Museum, *Capturing our Heritage, Scottsville Floods,* accessed July 15, 2021, https://scottsvillemuseum.com/floods/home.html.

2. NOAA, *Preliminary Reports*, 112-113.

3. NOAA, *Final Report*, 6.

4. Scottsville Museum, *Scottsville Floods*. Stage was determined after the flood by measuring the mud line on the piers of a bridge over the James River.

5. Ibid.

6. NOAA, *Preliminary Reports*, 113.

7. James Berry, *The Richmond Flood* (Lubbuck, Texas: C.F. Boone, 1972), 5.

8. Ibid., 25.

9. NOAA, *Final Report*, 8. "Precautionary measures in Richmond were extensive and extremely well executed."

10. National Trust for Historic Preservation, *Shockoe Bottom*, 2019, accessed July 15, 2021, https://savingplaces.org/places/shockoe-bottom.

11. James Berry, *The Richmond Flood*, 7.

12. NOAA, *Preliminary Reports*, 112-114.

13. "Hard-Hit Richmond Takes Flood Calmly," *Washington Post*, June 25, 1972, A-21.

14. National Trust for Historic Preservation: *Shockoe Bottom*.

15. *James River Flood, 1972,* accessed July 15, 2021, https://uvadisasters.fandom.com/wiki/James_River_Flood,_1972.

16. In August, 2004, the remnant of Hurricane Gaston brought locally intense rain to the Richmond area. Flash flooding on Shockoe and Giles Creeks rushed through Shockoe Bottom, again inundating homes and businesses. Ironically, the flood wall built to hold back the James impeded drainage from the creeks behind the wall. "Why, Richmond, Why? Officials say flood wall didn't fail during Gaston in 2004," *Times-Dispatch*, Richmond, Virginia, October 2, 2015: "The wall was built to hold the river," a city official said, "but the water didn't come from the river, it came from the sky."

THE OPERATORS

1. "Agnes Spreads Terror, Calms Down in Georgia," *The Evening News*, Harrisburg, Pennsylvania, June 20, 1972, 1.

2. "The Anatomy of a Disaster," *Harrisburg Patriot-News,* special edition, June 28, 1972.

3. John Baer, "Flood Worse, Times Better," *Philadelphia Daily News*, September 12, 2011.

4. "The IBM 1130 Computing System," accessed June 30. 2021, https://www. ibm.com/ibm/history/exhibits/1130/1130_intro.html.

PILGRIMAGE

1. "Heavy Rain Due Today," *Intelligencer-Journal*, Lancaster, Pennsylvania, June 21, 1972, 1.

2. The destruction of Camp Davis is fully told at Mary A. Shafer, *Devastation on the Delaware: Stories and Images of the Deadly Flood of 1955* (Riegelsville, Pennsylvania:

Word Forge Books, 2015); also, Frank Whelan, "Hurricane Diane Proved Tragic for Campers," *The Morning Call*, Allentown, Pennsylvania, October 13, 2004.

3. White's remarks at the East Stroudsburg conference were reported to the author by hydrologist Mike Gwinner.

SUPERSTORM

1. DeAngelis and Hodge. *Preliminary Climatic Data Report*, 6.

2. National Hurricane Center Director Robert Simpson detailed the meteorological processes by which Agnes combined with the stalled midwestern cold front to form a monster storm:

    > *Agnes was reduced to depression strength as it turned northeastward into Georgia on the 20th. However, as a ridge of high pressure over the western Atlantic continued to build, a major extra-tropical trough approached the weakening depression, Agnes' circulation experienced a renewed acceleration as a result of a release of fresh energy from baroclinic sources, and the central pressure again fell. Early on the 21st, a secondary Low center developed west of the old Agnes depression center. This complex system again reached tropical storm strength late on the 21st while the centers were still over land in North Carolina. The secondary Low center remained inland and ultimately became dominant, but not before the Agnes center moved offshore near Norfolk, VA, that evening to almost regain its previous strength. ... The two centers turned northward and westward under the steering influence of a new cold low that had developed over the Ohio valley. ...*

    Simpson and Hebert, *Atlantic Hurricane Season of 1972*, 6.

3. 2012's "Superstorm" Sandy, and the devastating flooding from 2011's Hurricane Lee, had meteorological similarities to Agnes. All three were remnant hurricanes enhanced by extra-tropical low pressure systems advancing from the Midwest.

4. Simpson and Hebert, *Atlantic Hurricane Season of 1972*, 9: *"Not as well forecast was the re-intensification of the storm center while over land, the increased intensity of rainfall, and attendant record-breaking floods."*

5. NOAA, *Preliminary Reports*, 122-126. Transcripts of NWS bulletins for most stations throughout the storm are reported.

6. NOAA, *Final Report*, 9.

7. "Flashback to when a hurricane similar to Harvey hit DC," WUSA-9 News, Washington, DC, August 28, 2017, accessed Oct 14, 2019, https://www.wusa9.com/article/weather/flashback-to-when-a-hurricane-similar-toharvey-hit-dc/468706991.

8. "Flooding in Area Kills 10, Washes out Road, Bridges," *Washington Post*, June 23, 1972, A-1.

9. "12,000 Evacuate Dwellings; Roads, Bridges Out, Water Supply Periled," *Washington Post*, June 23, 1972, A-1; 16; "N. Virginia has Water Crisis in Flood Wake", *Washington Post*, June 24, 1972, A-16.

10. "Flooding in Area Kills 10, Washes out Road, Bridges," *Washington Post*, June 23, 1972, p. A-1.

LONE WOLF

1. U.S. Army Corps of Engineers, Baltimore District, *Tropical Storm Agnes, June 1972: Post Flood Report* (two volumes), 1974, E-24.

2. "40 Years after Agnes, Officials say Laurel is Better Prepared for Storms," *Baltimore Sun*, June 20, 2012.
3. Rock Creek crested at 16.2 ft, with a discharge of 12,500 cfs, on June 22, 1972, records that have never since been approached.
4. Paul Vitello, "Theodore Reed, Who Lifted National Zoo's Profile, Dies at 90," *New York Times*, July 7, 2013.
5. Carrie McCloud, "Today in History: Hurricane Agnes Floods National Zoo," *Washington City Paper*, June 22, 2011.
6. Mike Morgan, "The Other Panda," *Smithsonian News Service*, reported in *The Morning Call*, Allentown, Pennsylvania, August 11, 1987.
7. "Agnes: 3-Day Flood, 20-Year Wake," *Washington Post*, Washington, DC, July 5, 1992, D-9.
8. "Water Even Breached White House Security," *Evening Star*, Washington, DC, June 22, 1972, B-3.
9. "Wide-Ranging Floods Stagger D.C. Area," *Washington Post*, June 24, 1972, A-1.
10. "Agnes Leaves Devastation," *Washington Post*, June 24, 1972, A-1.

EAGLE SCOUT

1. Ellicott City suffered disastrous flash floods in 2016 and 2018. Several people were killed when persistent local thunderstorms supercharged little creeks that wind down from the hillsides through culverts under the central business district. Allen H. Kittlemann, *Ellicott City Flood Mitigation Plan*, Howard County Council, 2018; Also National Weather Service, *May 27th, 2018 Flooding at Ellicott City & Catonsville, MD*. July 2, 2018, accessed June 20, 2021, https://www.weather.gov/lwx/EllicottCityFlood2018.
2. David Healey, *Great Storms of the Chesapeake* (Charleston, SC, The History Press, 2012), 50-54.
3. Kevin Ambrose, "Beyond Rescue: Ellicott City's Bizarre, Rainless Flood and its Deadly 20-foot Wall of Water," *Washington Post*, May 29, 2018.
4. Mike Bowler, "Flights over Floodwater Reveal Extent of Tragedy in State," *Baltimore Sun*, June 23, 1972, 2.
5. *The Flood of 1972* (Ellicott City, Maryland: The Times Newspapers, 1972).
6. "Swollen Patapsco caught 4 in its Path; 1 Survived," *Baltimore Sun*, Baltimore, Maryland, June 24, 1972, A-8.
7. "Forty Years Later, Agnes Remains a Benchmark for County Disasters," *Baltimore Sun*, June 20, 2012, accessed March 2, 2021, https://www.baltimoresun.com/ph-ho-cf-agnes-anniversary-0621-20120620-story.html.
8. Ibid.
9. The previous record of 6,250 cfs was set in 1952. *The Flood of 1972* (Ellicott City).
10. "Surprise , then Panic and Heroism," *Baltimore Sun*, Baltimore, Maryland, June 22, 1972, 1.
11. *The Flood of 1972* (Ellicott City).

I CAN'T SWIM

1. "Patriot-News Cancels Edition for First Time," *Patriot/Evening News, Special Flood Edition*, Harrisburg, Pennsylvania, June 28, 1972, 2.
2. John Baer, "Flood Worse, Times Better," *The Inquirer*, Philadelphia, Pennsylvania,

Sept. 12, 2011; Dan Cupper, "The Agony of Agnes," *Aspire Magazine*, June 1992; "Six Dead in County, *Patriot/Evening News, Special Flood Edition*, Harrisburg, Pennsylvania, June 28, 1972, 1; NOAA, *Final Report*, 16.

3. 26.36 feet, about two feet higher than 1933.

4. "Five Dead, Missing," *Daily Record*, York, Pennsylvania, June 23, 1972, 1; "Saga of Evacuation along the Codorus." *Daily Record*, York, Pennsylvania, June 29, 1972; "Tropical Storm Agnes arrived in York this Week in 1972, Bringing over a Foot of Rain," *Daily Record*, York, Pennsylvania, June 18, 2019.

5. "Six Covered Bridges Lost," *Intelligencer - Journal*, Lancaster, Pennsylvania, June 23, 1972, 1.

6. "Amish Man, Son Lost in Stream," *Intelligencer-Journal*, Lancaster, Pennsylvania, June 23, 1972; "Floods Didn't Pick Victims," *Pocono Record*, Stroudsburg, Pennsylvania, June 24, 1972, 1.

7. The Brandywine River reached a record stage of 16.5 feet on the USGS gauge during Agnes. A new record of 21 feet was set when the remnant of Hurricane Ida dumped 8+ inches of rain in the Brandywine watershed on September 1, 2021

8. "Wyeth Museum Flooded," *Evening Sun*, Baltimore, Maryland, June 23, 1972, 1.

RATS!

1. "Rats Invade Shapp Mansion Ruins," *The Sentinel*, Carlisle, Pennsylvania, June 28, 1972; "Swollen River Encircles Governor's New Mansion," *Intelligencer Journal*, Lancaster, PA, June 24, 1972; "Shapps Share Misery – Mansion Under Water," *Philadelphia Inquirer*, June 24, 1972.

2. "Come Inside the Governor's Mansion," *Morning Call*, Allentown, Pennsylvania, April 20, 2007.

3. "Shapps Will Live in State Mansion," *The News-Item*, Shamokin, Pennsylvania, December 9, 1970.

4. "Swollen River Encircles Governor's New Mansion," *Intelligencer Journal*, Lancaster, PA, June 24, 1972, 9.

5. "Shapps Share Misery – Mansion Under Water," *Philadelphia Inquirer*, June 24, 1972.

6. E.T. Engman, L.H. Parmele and W.J. Gburek, "Hydrologic Impact of Tropical Storm Agnes," *Journal of Hydrology* 22, Issues 1-2 (June 1974): 179-193,

7. "Swollen River Encircles Governor's New Mansion," *Intelligencer Journal*, Lancaster, PA, June 24, 1972, 9.

8. "Shapp Rescues his Dog from Flooded Mansion," *Philadelphia Inquirer*, June 25, 1972, 16.

9. "Swollen River Encircles Governor's New Mansion," *Intelligencer Journal*, Lancaster, PA, June 24, 1972, 9.

10. "Come Inside the Governor's Mansion," *Morning Call*, Allentown, Pennsylvania, April 20, 2007.

11. "Neighboring YMCAs had Trouble with Agnes," *The News*, Frederick, Maryland, July 15, 1972.

12. "Two Escape as Rail Span Collapses," *Daily Item*, Sunbury, Pennsylvania, June 22, 1972, 1.

13. "Special Issue: The Flood," *Penn Central Post* (Philadelphia: Penn Central Railroad Historical Society, August-September 1972).

BUCKET BRIGADE
1. "River on Rise Again", *Philadelphia Inquirer*, June 24, 1972, 4.
2. U.S. Environmental Protection Agency, Division of Oil and Hazardous Materials, *On-Scene Coordinator's Report: Schuylkill Oil Spill II, June-October 1972*, March, 1974: 8.
3. "Berks Associates Founder Roy H. Schurr Dies at 81," *The Mercury*, Pottstown, Pennsylvania, December 24, 1970.
4. "Flood Hits Townships, Dam in Area Breaks," *The Mercury*, Pottstown, Pennsylvania, June 23, 1972, 15.
5. EPA, *On-Scene Coordinator's Report*, 13.
6. Ibid., 13.
7. Ibid., 13-14.
8. U.S. Environmental Protection Agency, Office of Water Program Operations, *Environmental Effects of Schuylkill Oil Spill II*, 1975.
9. "Berks Oil Spill Set at 8 Million Gallons," *The Mercury*, Pottstown, Pennsylvania, June 29, 1972, p. 1.
10. "Losses Heavy, Cleanup Hampered," *The Mercury*, Pottstown, Pennsylvania, June 27, 1972.
11. "Coast Guard and EPA Start Oil Cleanup on Schuylkill," *The Mercury*, Pottstown, Pennsylvania, June 30, 1972, 9.
12. "Federal Environmental Chief Examines Oil Spill Mess," *The Mercury*, Pottstown, Pennsylvania, July 1, 1972, 1.
13. "Bucket Brigade Cleans Up Oil Along the Schuylkill," *New York Times*, July 7, 1972: 13.
14. Ibid.
15. EPA, *On-Scene Coordinator's Report*, 69.
16. "Flooding Caused Biggest Inland Oil Spill in U.S.," *Philadelphia Inquirer*, June 29, 1972, 4;"Oil Spill Makes History," *Republican and Herald*, Pottsville, Pennsylvania, July 1, 1972, 2;
    National Oceans and Atmospheric Administration, Office of Response and Restoration. *Largest Oil Spills Affecting U.S. Waters Since 1969*, accessed Dec. 4, 2019, https://response.restoration.noaa.gov/oil-and-chemical-spills/oil-spills/largest-oil-spills-affecting-us-waters-1969.html.
    By comparison, 11 million gallons of crude oil spilled in Alaska's 1989 Exxon Valdez maritime disaster.

MAN OVERBOARD
1. "The Schuylkill; A Writhing, 100-mile-long Monster," *Philadelphia Inquirer*, June 24, 1972, 3.
2. NOAA, *Final Report*, 14.
3. Only once since Agnes, in 2006, has the Schuylkill River at Pottstown risen as high as 20 feet, 9 feet lower than the Agnes record.
4. "Industry in Pottstown Crippled by Schuylkill," *The Mercury*, Pottstown, PA, June 23, 1972, 1.
5. "Policeman Worries, Aids Elderly Women," *The Mercury*, Pottstown, PA, June 24, 1972, 13.

6. "Schuylkill Searched for Missing Officer," *Philadelphia Inquirer*, June 24, 1972, 5; Cop Pulled from Schuylkill after Floating 3 Miles," *Philadelphia Inquirer*, June 23, 1972; "Agnes – And the Great Flood of '72", *Sunday Bulletin, Special Issue*, Philadelphia, Pennsylvania, 1972.

7. "Policeman's Body Found in Schuylkill," *Philadelphia Inquirer*, June 28, 1972, 4.

8. "Dying Hurricane Gives Sunday Punch," *The Mercury*, Pottstown, Pennsylvania, June 23, 1972, 26: "Drums containing highly flammable chemicals were floating down the Schuylkill River."

   Also, U.S. Department of Commerce, National Oceanic and Atmospheric Administration, *Incident News:Schuylkill River Spill*, accessed Sept. 30, 2021, https://incidentnews.noaa.gov/incident/6214#!513897.

9. "800 City Employees in Flood Cleanup," *Philadelphia Inquirer*, June 25, 1972, 19.

EPICENTER

1. In 1972 the Genesee River was not within the purview of any RFC; predictions for the Genesee were handled by WSO Rochester. NOAA, *Final Report*, 14.

2. "Weather Watch," *Democrat and Chronicle*, Rochester, New York, June 21, 1972: 22.

3. NWS Meteorologist George Shielein, acting MIC at WSO Rochester during Agnes, cited in "When Two Storms Got Together," *Democrat and Chronicle*, Rochester, New York, June 24, 1972: 2;

   Also, NWS Meteorologist Dave Nicosia, WSO Binghamton, in Frey, Brian (Writer and Producer), 2012, *The Flood of '72*. Binghamton, New York: WSKG Public Telecommunications Council.

4. Diane Ryan, *The Great Flood of '72 Revisited*, Almond (New York) Historical Society, 2007, accessed February 15, 2020, http://www.usgennet.org/usa/ny/town/almond/Archives%20List/Flood72/Flood72Revisited.htm.

5. "Storm, Offshoot of Hurricane Agnes, Centers over Wellsville for 14 Hours," *Daily Reporter*, Wellsville, New York, June 22, 1972: 1.

6. Ibid.

7. "City Slept at Time Storm Warnings Issued," *The Daily News*, Lebanon, Pennsylvania, July 3, 1972, 10; "Weathermen 'Lost' Agnes," *Evening Star*, Washington, DC, June 30, 1972, citing NOAA administrator Robert White.

8. NWS meteorologist Dave Nicosia, in Frey, *The Flood of '72*.

9. "When 2 Storms Got Together ... Rivers, Streams were Already Full," *Democrat and Chronicle*, Rochester New York, June 24, 1972, p. 2.

   Many Ph.D. dissertations and other scientific studies were published to explain the extraordinary re-development of Agnes, its merger with the low-pressure system along the stalled cold front, its turn westward over the Twin Tiers, and the prodigious rains it produced. *e.g.*, Geoffrey DiMego and L.F. Bosart, "The Transformation of Tropical Storm Agnes into an Extra-tropical Cyclone," *Monthly Weather Review* 110, 1982: 385-411.

10. The maximum total rainfall in the Twin Tiers region was 16 inches, recorded at Andover in Allegany County, New York. J.F. Bailey, J.L. Patterson, and J.L.H. Paulhus, *Hurricane Agnes Rainfall and Floods June-July 1972*, Geological Survey Professional Paper 924 (Washington, DC: U.S. Government Printing Office, 1975), 47.

11. "Second Flood Friday Adds to Damage in Area," *Daily Reporter*, Wellsville, New York, June 24, 1972, 1.

12. Frey, *The Flood of '72.*

13. NOAA, *Final Report*, 23: "Radar reports ... did not reflect the heavy rainfall rates during this event."

GRAND CANYON OF THE EAST

1. NOAA, *Final Report*, 14. The Genesee River has since been put under the umbrella of the Northeast RFC, today headquartered in Norton, Massachusetts, covering New England and parts of New York.

2. Ibid., 15.

3. Ibid., 14.

4. Acting MIC George Sheinlein, cited in "Storm, Offshoot of Hurricane Agnes, Centers over Wellsville for 14 Hours," *Daily Reporter*, Wellsville, New York, June 22, 1972: 1.

5. "Storm, Offshoot of Hurricane Agnes, Centers over Wellsville for 14 Hours," *Daily Reporter*, Wellsville, New York, June 22, 1972, 1; "When 2 Storms Got Together ... Rivers, Streams were Already Full," *Democrat and Chronicle*, Rochester, New York, June 24, 1972: 2.

6. Ibid., 2.

7. The gauges were destroyed in the flood, so maximum stage and discharge were determined by mud lines on nearby structures.

8. "Storm, Offshoot of Hurricane Agnes, Centers over Wellsville for 14 Hours," *Daily Reporter*, Wellsville, New York, June 22, 1972, 1.

9. "Three presumed dead," *Daily Reporter*, Wellsville, New York, June 22, 1972, 1.

10. "When 2 Storms Got Together... Rivers, Streams were Already Full," *Democrat and Chronicle*, Rochester, New York, June 24, 1972, 2.

11. United States Department of Commerce, National Weather Service, "Historic Flood June 1972 – Hurricane Agnes and the Genesee River Flooding," accessed April 18, 2020, *https://www.weather.gov/nerfc/hf_june_1972*: "Both of the official river gauges – at Scio and Wellsville – were destroyed by the flooding."

12. "Hospital Staff was Prepared for Collapse," *Daily Reporter*, Wellsville, New York, June 24, 1972, 4.

13. "Second Flood Friday Adds to Damage in Area," *Daily Reporter*, Wellsville, New York, June 24, 1972, 1.

14. Tom Cook and Tom Breslin. *Pieces of the Past – Artifacts, Documents, and Primary Sources from Letchworth Park History*, accessed April 23, 2020, *http://www.letchworthparkhistory.com/flood.html*.

15. "Top Flood Level Due Today," *Democrat and Chronicle*, Rochester, New York, June 24, 1972: 8.

16. United States Army Corps of Engineers, Buffalo District, *History of Mount Morris Dam*, accessed April 23, 2020, https://www.lrb.usace.army.mil/Missions/Recreation/Mount-Morris-Dam/Project-History/.

    Also, Luman F. Robison, *Floods in New York, 1972, with Special Reference to Tropical Storm Agnes*, United States Geological Survey, Water Resources Division. Albany, New York, 1976, 11.

17. "Floods Leave Counties Reeling," *Democrat and Chronicle*, Rochester, New York, June 24, 1972, 8.

18. "Rochester Spared as State digs out from Devastation," *Democrat and Chronicle*, Rochester, New York, June 25, 1972: 1.

19. Top Flood Level Due Today," *Democrat and Chronicle*, Rochester, New York, June 24, 1972: 8.

20. $1.28 billion in 2020 dollars. Corps of Engineers. "History of Mount Morris Dam"; Chris Clemens, "How the Mount Morris Dam Saves Rochester," *Exploring Upstate, 2019*, accessed April 24, 2020, https://exploringupstate.com/mt-morris-dam/.

LAKE OF SALVATION

1. Paul Lamont and Scott Saggett, writers, director and producer. Lake of Betrayal, Castle Films, 2017, accessed September 30, 2021, https:// www.lakeofbetrayal. com/.

2. Nathan C. Grover, *The Floods of 1936*. United States Geological Survey, Water Supply Paper 799 (Washington, DC: U.S Government Printing Office, 1937), 35; Shank, *Great Floods of Pennsylvania*, 44.

3. The Ohio RFC is now (2020) located at Wilmington, Ohio.

4. NOAA, *Final Report*, 24.

5. "City Slept at Time Storm Warnings Issued," *Daily News*, Lebanon, Pennsylvania, July 3, 1972, 10.

6. NOAA, *Final Report*, 24.

7. "City Slept at Time Storm Warnings Issued," *Daily News*, Lebanon, Pennsylvania, July 3, 1972,10; NOAA, *Final Report*, 23.

8. "Flood of 1972 Remembered," *Times-Herald*, Olean, New York, June 24, 2012.

9. NOAA, Final Report, 24; *Flood: The Southern Tier's June 1972 Disaster: A Pictorial Review* (Hornell, New York: W.H. Greenhow Co., 1972), 6.

10. "Flood of 1972 Remembered," *Times Herald*, Olean, New York, June 24, 2012.

11. "Flooding in Area Above Kinzua Dam," *Times-Mirror and Observer*, Warren, Pennsylvania, June 23, 1972: 1.

12. "The Rains Came ... And the Floods went Up," *Times-Observer,* Warren, Pennsylvania, March 10, 2018.

13. "Kinzua Dam Saved Warren Area $1,290,000 Damage from Flood," *Times-Mirror and Observer,* Warren, Pennsylvania, August 30, 1972, 12.

14. "Bulk of Rainfall Hit Unprotected River Basins," *Post-Gazette*, Pittsburgh, Pennsylvania, June 24, 1972, 4; NOAA, Final Report, 25.

15. NOAA, *Final Report*, 25.

16. "Heinz Hits Crisis Reporting System," *Post-Gazette*, Pittsburgh, Pennsylvania, June 30, 1972, 6.

17. NOAA, *Final Report*, 23

18. "City Slept At Time Storm Warnings were Issued," *Daily News*, Lebanon, Pennsylvania, July 3, 1972, 10.

19. "Three Rivers Waters Wetting Pittsburgh, *"Times-Mirror and Observer*, Warren, Pennsylvania, June 24, 1972, 1.

20. "When Hurricane Agnes Slammed Soggy Pittsburgh," comment of Daniel Kablack in *Archives of the Pittsburgh Post-Gazette*, July 15, 2015, accessed May 13, 2020, https://newsinteractive.post-gazette.com/thedigs/2015/07/15/when-hurricane-agnes-slammed-soggy-pittsburgh/.

21. "Agnes Turn-About Dealt $45 Million Blow Here," *Pittsburgh Press*, Pittsburgh, Pennsylvania, June 25, 1972, 1.

22. Ibid.; "Heinz Hits Crisis Reporting System," *Post-Gazette*, Pittsburgh, Pennsylvania, June 30, 1972, 6.

23. Ibid.

24. "Pittsburgh Struggles to Limit Flood Loss," *Post-Gazette*, Pittsburgh, Pennsylvania, June 24, 1972, 4.

25. "Alice Cooper Show Almost Off," *Post-Gazette*, Pittsburgh, Pennsylvania, July 12, 1972.

26. "Pittsburgh Struggles to Limit Flood Loss," *Post-Gazette*, Pittsburgh, Pennsylvania, June 24, 1972, 4.

27. "When Hurricane Agnes Slammed Soggy Pittsburgh," Comment of Bob Suchy in *Archives of the Pittsburgh Post-Gazette*, July 15, 2015, accessed May 13, 2020, https://newsinteractive.post-gazette.com/thedigs/2015/07/15/when-hurricane-agnes-slammed-soggy-pittsburgh/.

28. Agnes Turn-About Dealt $45 Million Blow Here," *Pittsburgh Press*, Pittsburgh, Pennsylvania, June 25, 1972, 1.

29. "Clean-up Begins as Flood Waters Recede in East," *Baltimore Sun*, June 26, 1972, A-1.

30. "Heinz Hits Crisis Reporting System," *Post-Gazette,* Pittsburgh, Pennsylvania, June 30, 1972, 6: "Long was praised by [NOAA director] Robert White for staying at his post for six days and eight nights."

31. "8 Dams Prevented Record Flood," *Pittsburgh Press*, June 25, 1972, 20.

32. $6.13 billion in 2020. "When Hurricane Agnes Slammed Soggy Pittsburgh," *Archives of the Pittsburgh Post-Gazette*, July 15, 2015, accessed May 13, 2020, https://newsinteractive.post-gazette.com/thedigs/2015/07/15/when-hurricane-agnes-slammed-soggy-pittsburgh/.

THE FIREMAN
1. "Press Hunt for Savona Man Believed Drowned," *The Leader*, Corning, New York, June 22, 1972, 2.

2. Nicholas R. Hoye, Michael R. Orr and Joseph A. Anastasi, *The Flood and the Community* (Corning, New York: Corning Glass Works, 1976).

   Also, Frey, *The Flood of '72.*

3. After the Agnes floods the levee system at Gang Mills was expanded and strengthened. New York Department of Environmental Conservation, *Gang Mills Flood Damage Reduction Project*, accessed Jan. 10, 2020, https://www.dec.ny.gov/docs/water_pdf/fcpprjgngmls.pdf.

4. Corps of Engineers, *Post Flood Report*, Vol. II, A-67-69.

5. Hoye, *The Flood and the Community*, 22-25.

6. Ibid, 26-27.

THE MAYOR'S BROADCAST
1. Frey, *The Flood of '72.*

2. New York Department of Environmental Conservation, "Corning and Painted Post Flood Damage Reduction Project," accessed Jan 10, 2020, *https://www.dec.ny.gov/docs/water_pdf/fcpprjcornpp.pdf.*

3. Hoye, *The Flood and the Community*, 18.

4. "Rivers Rise, Rain May Continue Friday," *The Leader*, Corning, New York, June 22, 1972, 3.

5. Hoye, *The Flood and the Community*, 29.

6. "Sturdy Dikes Save Corning Once More," *The Leader*, Corning, New York, June 22, 1972, 3.

7. Frey, *The Flood of '72*: "We were listening to the radio, and the mayor came on and said the dikes would hold, and there was nothing to worry about. So we went to bed that evening thinking all was safe and sound."

8. Bragg, Meridith (producer). *Stories from the Flood of '72: John Fox*, Corning Museum of Glass, 2012, accessed June 20, 2021, https://www.cmog.org/video/stories-flood-72-john-fox; Hoye, et. al., *The Flood and the Community*, 23, 42.

9. "No One Knew Flood Would Hit," *The Leader*, Corning, New York, June 28, 1972, 1.

10. "Drama Unfolds in First Hours," *Star-Gazette*, Elmira, New York, July 2, 1972.

11. Hoye, et. at., *The Flood and the Community*, 33.

12. Ibid.

SHATTERED GLASS

1. Hoye, *The Flood and the Community*, 41.

2. Ibid., 54-55.

3. Charleen Edwards, *Museum Under Water* (Corning, New York: Corning Museum of Glass, 1977).

4. Frey, *The Flood of '72*.

5. "Accused of Looting After they Saved 60," *The Leader*, Corning, New York, June 28, 1972, 4.

6. "Hard-Hit Leader Moving Back About August 1," *The Leader*, Corning, New York, July 7, 1972, 3.

7. "No Snakes," *The Leader*, Corning, New York, July 3, 1972, 2.

8. "Special Edition," *The Leader and Elmira Star-Gazette*, Corning/Elmira, New York, June 24, 1972.

9. "Hard-Hit Leader Moving Back About August," *The Leader*, Corning, New York, July 7, 1972, 3.

10. Bragg, Meridith, producer, *Stories from the Flood of '72*.

11. Reported to the author by Dr. Robert Brill.

12. Corps of Engineers, *Post Flood Report*, Vol. II, A-99.

13. Blanchard, Wayne. *Deadliest Natural Disasters*.

14. "No One Knew Flood Would Hit," *The Leader*, Corning, New York, June 28, 1972,

DERENZO'S PLAN

1. Episodes from the flood at Elmira are largely drawn from:
   Thomas Byrne, "The Impossible Flood," *Chemung County Historical Journal* (Elmira, New York, 1972);
   Kirk W. House, *The 1972 Flood in New York's Southern Tier* (Charleston, South Carolina: Arcadia Publishing, 2012);
   Marvin W. Copp, *Floods of the Chemung Watershed 1794-1972: A Day to Remember*, June 23, 1972 (Golos Publishing: Elmira Heights, NewYork, 1975);
   Kenneth I. Darmer and Lloyd A. Wagner, *Flood of June, 1972, at Elmira, New York.*

United States Geological Survey, Hydrologic Investigations Atlas HA-518, 1973;

Frey, *The Flood of '72*;

Rich LaVere (writer and producer), *Facing the Wall* (LaVere Media, Elmira, New York, 2010);

Robison, *Floods in New York, 1972*.

2. In 1972 the nearest USGS river gauge was at the village of Chemung, five miles downstream from Elmira. There was a gauge at the Lake Street Bridge at Elmira for reference by city authorities, but not included in the calculus of river forecasts. A USGS gauge was installed at Elmira in 1988. Corps of Engineers, *Post Flood Report*, Vol. II, A-108.

3. "The Engineers' Warnings were Ignored: They Knew Disaster was Waiting," *Star-Gazette*, Elmira, New York, July 2, 1972, 25.

4. Frey, *The Flood of '72*.

5. Ibid.

6. "Forecast: A Chance of Showers," *Sunday Telegram*, Elmira, New York, July 2, 1972, 25.

7. "It's a Disaster – But City was Prepared for It," *Star-Gazette*, Elmira, New York, June 23, 1972, 7.

8. "Some Didn't Believe; They do Now," *Star-Gazette*, Elmira, New York, July 2, 1972, 29.

9. "Forecast: 'A Chance of Showers'," *Sunday Telegram*, Elmira, New York, July 2, 1972, 25.

10. Byrne, "Impossible Flood," 2150.

11. *Flood: The Southern Tier's June 1972 Disaster: A Pictorial Review*. Hornell, New York: W.H. Greenhow Co., 1972; "Past Critical Stage in Big Flats Area," *The Leader*, Corning, New York, June 30, 1972, 1; Byrne, *The Impossible Flood*, 1056. Other reports pegged the spill at 3 million gallons "of oil," *e.g.*,"Flood Watch on Again After Rains," *The Leader*, Corning, New York, June 30, 1972, 17.

12. "Forecast: A Chance of Showers," *Sunday Telegram*, Elmira, New York, July 2, 1972, 30.

13. "It's a Disaster – But City was Prepared for It," *Star-Gazette*, Elmira, New York, June 23, 1972, 7.

14. "Devastation – Windows Smashed, Stores Ripped Apart, Slime, Mud," *Sunday Telegram*, Elmira, New York, June 25, 1972, 1.

15. "Some Didn't Believe; They do Now," *Star-Gazette*, Elmira, New York, July 2, 1972, 29.

16. "Ham Operators: Messengers for the Distressed," *Star-Gazette*, Elmira, New York, June 25, 1972, 5.

17. "Southport: We're Like a Lost Nation over Here," *Star-Gazette*, Elmira, New York, June 28, 1972, 16.

18. "Guard Choppers come to the Rescue," *Star-Gazette*, Elmira, New York, June 24, 1972, 2.

19. Darmer, et. al., *Flood of June, 1972, at Elmira, New York*.

20. "Half of City Evacuated," *Star-Gazette*, Elmira, New York, June 23, 1972, 1.

21. About $1.6 billion in 2020 dollars. Corps of Engineers, *Post Flood Report*, Vol II, A-109.

22. John and Marina Nickerson, *It Sprinkled, it Rained, and it Poured* (music, Elmira, New York: WENY Radio, 1972), video by Erin Doane, producer (Elmira, New York: Chemung County Historical Society, 2012), accessed April 9, 2020, https://www.youtube.com/watch?v=q7Q9xHjcv_g.

THE RELAY

1. Frey, *The Flood of '72*.

2. "Broome Observers Relayed Flood Data," *Press and Sun-Bulletin*, Binghamton, New York, June 29, 1972, 9. These events were also reported to the author by hydrologist Mike Gwinner of MARFC and Marsha Root Field, daughter of the late LaVern Root.

BOLD FORECAST

1. *After Agnes: A Triumph Over Destruction*. Wilkes-Barre, Pennsylvania: The Times-Leader, 1982, 6.

2. Mussari, Anthony J., *Appointment with Disaster* (Wilkes-Barre, Pennsylvania: Northeast Publishers, 1974), 55.

3. Townend's 109th National Guard regiment was attached to the 28th Division.

4. *After Agnes*, 6-7.

5. Mussari, *Appointment with Disaster*,59.

6. Ibid., 58.

7. Ibid., 61.

8. "34-foot Crest Expected before Susquehanna Falls," *Times Leader*, Wilkes-Barre, Pennsylvania, June 23, 1972.

9. Mussari, *Appointment with Disaster*, 12-13.

10. Robert G. Kiefer, director, *Storm – Hurricane Agnes/Wilkes-Barre*, Produced by Motion Picture Division, United States Department of Agriculture: A Public Information Film presented by Defense Civil Preparedness Agency, 1972.

11. *After Agnes*, 8.

12. Mussari, *Appointment with Disaster*, 66.

13. Kiefer, *Storm – Hurricane Agnes/Wilkes-Barre*.

14. Comment of Myron Gwinner, *The Great Flood of 1972, Wilkes-Barre is Dealt a Powerful Blow*, accessed February 12, 2022, https://petejoem.wordpress.com/2014/02/09/the-great-flood-of-1972-wilkes-barre-pa-is-dealt-a-powerful-blow/comment-page-/?unapproved=319&moderation-hash=1ca0f72dad85d74b95c930afdf820596#comment-319.

15. Albert Kachic, "Agnes Legacy", presented at Alert-FLOWS annual conference, Philadelphia, 2002; also, *National Weather Service Marks 25th Anniversary of Hurricane Agnes*, NOAA press release 97-R228, June 11, 1997.

16. Corps of Engineers, *Post Flood Report*, Vol I, 54.

17. National Advisory Committee on Oceans and Atmosphere, *The Agnes Floods: A Post-Audit of the Effectiveness of the Storm and Flood Warning System of the National Oceanic and Atmospheric Administration*, Nov. 22, 1972, 34.

18. NOAA, Final Report, 15: *Perhaps the most outstanding issuance of the whole disaster was a flood forecast sent by RFC Harrisburg to the civil defense office in Wilkes-Barre*

at 3:00 a.m. on Friday, June 23. ... This forecast triggered a mass evacuation of 80,000 to 100,000 persons, and is unquestionably responsible for preventing a disaster of unimaginable magnitude.

EVACUATE!

1. Mussari, *Appointment with Disaster*, 66.
2. Ibid., 16-17.
3. Ibid., 6-17.
4. Ibid., 17.
5. *Agnes Revisited*, a Special Report of the Times-Leader, Wilkes-Barre, Pennsylvania, June 18, 1992, 46-47.
6. Ibid., p. 46.
7. Ibid., p. 46.
8. "County Waiting Word from State on Health Department Formation," *Times Leader*, Wilkes-Barre, Pennsylvania, July 14, 1972, 27.
9. "Hurricane Agony," *Citizens Voice*, Wilkes-Barre, Pennsylvania. June 24, 2012.
10. Ibid.

SANDBAGS

1. "Dirt, Sand Needed to Hold Dike," *Times Leader/Evening News*, Wilkes-Barre, Pennsylvania, June 23, 1972, 3.
2. "State of Emergency Declared in Wilkes-Barre, Homes Evacuated," *Times Leader*, Wilkes-Barre, Pennsylvania, June 22, 1972, 2.
3. Jack Jarecki, "How the City Wasn't Saved," *Wyoming Observer*, Wyoming, Pennsylvania, July 2-8, 1972.
4. Kiefer, *Storm – Hurricane Agnes/Wilkes-Barre.*
5. Carl J. Romanelli, *The Wrath of Agnes* (Wilkes-Barre: Media Affiliates, Inc., 1972), 105.

OVER THE WALL

1. Bob Bradley, Bob, director, *Agnes: A Flood of Memories*. Film produced by Northeast Television Investors and WBRE-TV, 1992: "It took an agonizing 14 minutes before the sirens finally sounded."
2. "The Violent, Deadly Swath of Agnes," *Time Magazine*, July 3, 1972, 9.
3. "Sixth Drowning Victim Found in Yard in South Wilkes-Barre," *Times-Leader*, Wilkes-Barre, Pennsylvania, July 3, 1972, 6; "Sixth River Victim is Identified," *Times-Leader*, Wilkes-Barre, Pennsylvania, July 5, 1972, 1.
4. Bradley, *Agnes: A Flood of Memories.*
5. "Voices from Yesterday: I Feel Numb," *Agnes Revisited*, 53.
6. "State Police Among the First to Act," *Times-Leader*, Wilkes-Barre, Pennsylvania, July 5, 1972, 1.
7. "Drowning Victim's Funeral is Held," *Times-Leader*, Wilkes-Barre, Pennsylvania, June 28, 1972, 2.
8. "Flood Worker Dies in Water," *Times-Leader*, Wilkes-Barre, Pennsylvania, June 27, 1972, 7; Romanelli, *The Wrath of Agnes*, 23.
9. "The Night Wilkes-Barre's Dikes Failed," *Times-Leader*, Wilkes-Barre, Pennsylvania, July 1, 1972, 2.
10. Ibid.

11. "100 Animals Saved by Two Women of SPCA," *Times-Leader*, Wilkes-Barre, Pennsylvania, July 1, 1972, 1.

12. "Efforts Being Made to Save All Animals," *Times-Leader*, Wilkes-Barre, Pennsylvania, July 3, 1972, 2.

13. "Flood Stricken Pets also Get Help," *Daily Review*, Towanda, Pennsylvania, July 3, 1972; "Wide-Ranging Floods Stagger D.C. Area," *Washington Post*, June 24, 1972, A-1: One hundred head of cattle reported drowned in Prince William County, Virginia.

14. Corps of Engineers, *Post Flood Report*, Vol II, A-227 ($6.8 billion in 2020 dollars).

15. In his comprehensive review, Professor Wayne Blanchard counted three deaths in Luzerne county: Shock and Seiwell during the flood, and National Guardsman Robert Whitman during the cleanup (heart attack). Blanchard, *Deadliest Natural Disasters*.

   The author's review of contemporaneous accounts identifies seven deaths in Luzerne County, five during the flood and two more during the recovery:
   William Shock, of Wilkes-Barre, drowned June 23 during sandbag operations at the flood walls. "Sixth River Victim is Identified," Times-Leader, July 5, 1972, 1.
   William Seiwell, drowned June 23 after falling from a boat during rescue operations. "Flood Worker Dies In Water," Times-Leader, June 27, 1972, 7.
   John Metronick, 85, of Kingston, found in his home on Dawes Avenue June 26 or 27. "Five Drowning Deaths Recorded by Coroner and State Police," Times Leader, June 30, 1972, 12.
   Harald Henning, 73, of Kingston, found dead in his car "a victim of the flood". "Flood Victim's Funeral is Held," Times-Leader, July 3, 1972, 26.
   John Morris, 78, of Kingston, found dead in his home June 27, "a flood victim". "Kingston Man Victim of Flood," Times-Leader, July 3, 1972, 25.
   Marjorie Smith, 40, of Wilkes-Barre, July 1 of heart seizure while cleaning debris at her home. "Heart Seizure While Cleaning Debris Fatal," Times-Leader, July 3, 1972, 25.
   Sgt. Robert Whitman (National Guard), 42, of Johnstown, July 14 of a heart attack during emergency operations at Wilkes-Barre. "National Guard ... Honor Soldier Who Died on Duty," Times-Leader, July 17, 1972, 7.

16. NOAA, *Final Report*, 16. The entire excerpt reads:
   On behalf of the people of Pennsylvania and the Commonwealth Government, I extend to you and your fellow workers at the [River Forecast Center] our sincere thanks for the highly valued service so capably rendered in connection with last month's disastrous flood. ...
   Should there be any doubt as to the value of the forecast operations, I need cite only the forty-foot crest prediction for Wilkes-Barre, which came early Friday morning, June 23. Passed to the Luzerne County officials with a recommendation that everyone 'behind the dikes' be evacuated, that single bit of information unquestionably was responsible for the saving of countless human lives which otherwise would have been lost.

LOGS

1. Reported to the author by River Forecast Center hydrologist Joe Ostrowski.

2. This account of destruction of the Forty Fort Cemetery is primarily drawn from: Forty Fort Cemetery Association. *In Memoriam: June 23, 1972.* Memorial booklet published by the Association, October 31, 1988.

"Grim memories: Agnes cleanup included corpse recovery near cemetery,"
    Citizens Voice, Wilkes-Barre, Pennsylvania, June 22, 2018;
Bryan Glahn, *Hurricane Agnes in the Wyoming Valley*. (Charleston, South Carolina:
    Arcadia Publishing, 2017), 17-48.
3. *Agnes Revisited*, 39.

WE LOVE YOU

1. "Streams Rampage, Watch River Stage," *The Express*, Lock Haven, Pennsylvania,
    June 22, 1972, 1, 11.
2. "West Branch Seen Keeping within its Banks," *The Express*, Lock Haven,
    Pennsylvania, June 22, 1972,
3. "More Flood Observations," *The Express*, Lock Haven, Pennsylvania, June 26,
    1972, 2.
4. "Three Reportedly Dead at Lewisburg," *The Daily Item*, Sunbury, Pennsylvania,
    June 22, 1972, 1.
5. "Flood Victim's Body Recovered," *The Daily Item*, Sunbury, Pennsylvania, July
    10, 1972.
6. Susan Stranahan, *Susquehanna: River of Dreams* (Baltimore: The John Hopkins
    University Press, 1993), 133-134.
7. Corps of Engineers, Baltimore District, *Post Flood Report*, C-7-8.
8. "Morgan Declares Emergency," *Daily Item*, Sunbury, Pennsylvania, June 22, 1972, 4.
9. "The Fabridam Made it All Possible," *Daily Item*, Sunbury, Pennsylvania, June
    15, 1972.
10. "Floodwall Due U.S. Inspection," *Daily Item*, Sunbury, Pennsylvania, June 27,
    1972, 2.
11. "Susquehanna Crests at Record Levels, is Receding." *Daily Item*, Sunbury,
    Pennsylvania, June 24, 1972, 1.
12. Ibid.
13. "Floodwall Holding Despite Pressure," *Daily Item*, Sunbury, Pennsylvania, June
    25, 1972, 2.
14. Ibid.

DYNAMITE

1. John Paulson (writer and director) and Michael English (writer and executive
    producer), *Conowingo Dam: Power on the Susquehanna*, Maryland Public Television,
    2016.
2. "Conowingo Dam Visitors get Up-close Look at an Engineering and Construction
    Marvel," *The Aegis*, Bel Air, Maryland, Sept. 25, 2017.
3. Paulson and English, *Conowingo Dam*.
4. "Nixon Declares State a Disaster Area," *Baltimore Sun*, Baltimore, Maryland,
    June 24, 1972, 1.
5. This was a measure of "head", *i.e.*, the elevation of the reservoir surface in relation
    to the turbines in the power plant below.
6. "Nixon Declares State a Disaster Area," *Baltimore Sun*, Baltimore, Maryland,
    June 24, 1972, p. 1.
7. "Conowingo Dam Danger Passes as River Crests," *Baltimore Sun*, Baltimore,
    Maryland, June 25, 1972, 2.

8. "Remembering Agnes: We Really Took a Hit," *Baltimore Sun*, June 21, 2012.
9. Paulson and English, *Conowingo Dam*.
10. Ibid.
11. "Remembering Agnes: We Really Took a Hit," *Baltimore Sun*, June 21, 2012.
12. Evidently the Maryland Public Television film *Conowingo Dam: Power on the Susquehanna* was the first documented public telling of the story. Newspapers at the time reported only rumors.
13. "Remembering Agnes: We Really Took a Hit," *Baltimore Sun*, June 21, 2012.

LAST GASP
1. "More Rain in Area from Agnes," *Ottawa Journal*, Ottawa, Ontario, June 24, 1972, 5.
2. "This Month's Rain Worst Since 1943," *Ottawa Journal*, Ottawa, Ontario, June 26, 1972, 1.
3. "Man's House Vanished Right Before his Eyes," *Ottawa Citizen*, Ottawa, Ontario, June 26, 1972, 17.
4. "Freak Tornado Rips Maniwaki," *Ottawa Journal*, Ottawa, Ontario, June 26, 1972, 2.
5. "Aid Sought for Disaster Area," *Ottawa Citizen*, Ottawa, Ontario, June 26, 1972, 17.
6. "Freak Tornado Rips Maniwaki," *Ottawa Journal*, Ottawa, Ontario, June 26, 1972, 1-2.
7. "It was a Freak but a Vicious One," *Ottawa Citizen*, Ottawa, Ontario, June 26, 1972, 17.
8. Ibid.
9. Ibid.
10. "Aid Sought for Disaster Area," *Ottawa Citizen*, Ottawa, Ontario, June 26, 1972, 17.
11. "That Tree – That's What Saved Us," *The Gazette*, Montreal, Quebec, June 26, 1972, 1.
12. "Aid Sought for Disaster Area," *Ottawa Citizen*, Ottawa, Ontario, June 26, 1972, 17.
13. "Storm Mishaps Face Work Crews," *The Gazette*, Montreal, Quebec, June 23, 1972, 1; "Power Restored, Talks Set," *Ottawa Citizen*, Ottawa, Ontario, June 26, 1972, 5.
14. "Sons Drown after Fall," *Washington Post*, June 26, 1972, C-1; "Question of Life with Two Sons Dead," *Washington Post*, June 29, 1972, C-1, 5; NOAA, *Final Report*, 10.
15. Several other people died after the floods receded, *e.g.* a news crew when their helicopter crashed, and at least two people of heart attacks while cleaning up debris. Some authorities attribute these later fatalities to Agnes. Blanchard, Wayne. *Deadliest Natural Disasters.*
16. United States Department of Commerce, National Oceanic and Atmospheric Administration, *Mariners Weather Log* 16, no. 6 (June 1972), 378.
17. Ibid.

AFTERMATH
1. The full story of the years-long cleanup and recovery after the Agnes floods is beyond the scope of this work. The cleanup and recovery is a tale unto itself; see, *e.g.*:
Hoye, *et. al., The Flood and the Community*;
Mussari, *Appointment with Disaster*;
Timothy Kneeland, *Playing Politics with Natural Disaster: Hurricane Agnes, the*

1972 Election, and the Origins of FEMA (Ithaca, New York: Cornell University Press, 2020).

2. *Agnes Revisited,* 22; "Two Breweries Ship Water in Cans Instead of Beer," *Times Leader,* Wilkes-Barre, Pennsylvania, July 6, 1972, 15;
   David L. Krantz, *The Trouble with Agnes.* (Wilkes-Barre, Pennsylvania: Fowler, Dick & Walker, 1973), 94.

3. *Agnes Revisited,* 28.

4. Blanchard, *"Deadliest Natural Disasters."* Blanchard's review is supplemented by the author's research. Fatalities broken down by state are as follows:
   Florida: 9
   North Carolina: 2
   Virginia: 17
   Dist. of Columbia: 3
   Maryland: 21
   Delaware: 1
   New Jersey: 1
   Pennsylvania: 66
   West Virginia: 1
   New York: 26
   Total USA: 147
   Cuba: 16+
   Canada: 2

5. *After Agnes,* 20; "Up to 25,000 Still Homeless, CD Announces," *Times-Leader/ Evening News/Wilkes-Barre Record, Special Flood Edition,* June 30, 1972, 1.

6. "Tale of Two Storms, Agnes and Lee," *Citizens Voice,* Wilkes-Barre, Pennsylvania, June 16, 2012, accessed Feb 15, 2021, https://www.citizensvoice.com/news/tale-of-two-storms-agnes-and-lee/article_b6a2270a-d56f-511e-a2fa-340b883c3dd4.html;
   Paul K. Walker, *The Corps Responds: A History of the Susquehanna Engineer District and Tropical Storm Agnes,* United States Army Corps of Engineers, Baltimore District, 1976, 8.

7. "45 Years Later, Agnes Still on People's Minds," *The Weekender,* Wilkes-Barre, Pennsylvania, June 18, 2017.

8. Irene Fatis, contribution in *The Flood of '72 – Elmira Remembers,* accessed Dec. 31, 2020, https://www.elmira-ny.com/flood/index.shtml.

9. Ibid.

10. *Agnes Revisited,* 44.

11. Hoye, *The Flood and the Community,* 115-116.

12. "The Night Wilkes-Barre's Dikes Failed," *Times Leader,* Wilkes-Barre, Pennsylvania, July 1, 1972, 10.

13. "At W-B, Volunteers Carried the Day," *Times-Leader,* Wilkes-Barre, Pennsylvania, June 29, 1972, p. 1.

14. *Agnes Revisited,* 32.

15. Ibid, 29; Maj. Gen. Richard H. Groves, "The Agnes Disaster," *The Military Engineer* 65, no. 423 (January-February 1973), 15-21.

16. United States Congress. House. Committee on Banking, Finance and Urban

Affairs, Subcommittee on Policy Research and Insurance, *Aftermath of Hurricane Agnes*, Serial No. 101-147 (Washington: U.S. Government Printing Office, June 22, 1990), 45.

17. Kneeland, *Playing Politics*, 12-24.

18. The restoration of the Museum of Glass is well and fully told in:
Frey, *The Flood of '72*;
Hoye, *The Flood and the Community*;
John H. Martin, *The Corning Flood: Museum Under Water* (Corning, New York: Corning Museum of Glass, 1977).

19. Tommy Hilfiger, in Frey, *The Flood of '72*.

20. "Weather Bureau Probe is Asked," *News-Herald*, Panama City, Florida, June 21, 1972, 9.

21. "Tourist City to Sue Media; Cites 'Faulty" Agnes Reports," *Fort Lauderdale News*, Fort Lauderdale, Florida, June 27, 1972, 25.

22. "Weather Bureau Probe is Asked," *News-Herald*, Panama City, Florida. June 21, 1972, 9.

23. "Apalach Thanks Weather Bureau," *News-Herald*, Panama City, Florida, June 22, 1972, 1B.

24. "Heinz Faults Late Flood Warning," *Post-Gazette*, Pittsburgh, Pennsylvania, June 26, 1972, 8.

25. Ibid.

26. "Weathermen Recall Agnes' Surprise Turn," *Morning Herald*, Hagerstown, Maryland, June 30, 1972.

27. Kachic, *Agnes' Legacy*,

28. Kneeland, *Playing Politics*.

29. "Special Issue: The Flood," *Penn Central Post*, 1972

30. "Flood KO's Erie Railway," *Evening Star*, Washington, DC, June 27, 1972.

31. Among them, the Kueka Outlet Trail in New York's Finger Lakes region, the Northern Central trail from Baltimore to York, Pennsylvania, and the Lebanon Valley trail in central Pennsylvania.

32. *Luzerne County Flood Protection Authority*, accessed Dec. 31, 2020, http://www.lcfpa.org/history.html.

33. Jody A Schroath, "Dammed if you Do, Damned if you Don't," *Chesapeake Bay Magazine*, April/May, 2011.

34. Tom Horton, "Agnes is Still with Us," *Baltimore Sun*, June 27, 2012; "Body Blow for the Bay," *After the Deluge, Washington Post*, July 2, 2006, B-8.

35. Tom Horton, "Retrospective – The Damage Caused by Hurricane Agnes," *Washingtonian Magazine*, June 19, 2012.

36. Angus Phillips, "Storms Leave the Bay Cloudy, With the Future No Brighter," *Washington Post*, July 9, 2006, E-10.

37. Horton, "Retrospective."

38. A.M. Andersen, W.J. Davis, M.P. Lynch and J.R. Schubel,"Effects of Hurricane Agnes on the Environment and Organisms of Chesapeake Bay: Early Findings and Recommendations," *Special Reports in Applied Marine Science and Ocean*

Engineering, 29 (Virginia Institute of Marine Science, College of William and Mary, 1973), xiii, accessed Jan 7, 2021, https://doi.org/10.21220/V5M44V.

39. Len Lazarick, "Agnes' Historic Rainfall Taught Watershed a Concrete Lesson," *MarylandReporter.com,* June 27, 2012, accessed July 30, 2021, https://marylandreporter.com/2012/06/27/agnes-historic-rainfall-taught-watershed-a-concrete-lesson/.

EPILOG

1. "Commerce Honors NOAA Employees," *NOAA Magazine* no. 1, January 1974, 73.

2. United States Department of Commerce, *Gold Medal Citation,* October 1973.

3. United States Army Corps of Engineers, *Tropical Storm Agnes!* 1973.

4. Blaine P. Friedlander, "Hurricane Floyd was strong, but was no Agnes," *Cornell Chronicle* (Sept. 23, 1999). Cites Agnes as the "all time leader in precipitation."

5. Lisa Mathews, "Tropical Storm Lee: 29T Gallons of Water Create Flash Floods." *International Business Times* (Sept. 9, 2011). Cites Dr. Greg Forbes as source of rainfall calculation.

6. Christopher W. Landsea, "Hurricane Harvey's Rainfall and Global Warming," National Oceanic and Atmospheric Administration, accessed June 9, 2021, *https://www.aoml.noaa.gov/hrd/Landsea/harvey-global-warming.pdf.*

7. Angela Fritz and Jason Samanow (the "Capital Weather Gang"), "Harvey Unloaded 33 Trillion Gallons of Water in the U.S.," *Washington Post,* Sept 2, 2017.

8. "Hurricane Agnes: The Most Costly Storm," *Weatherwise* 25, no. 4 (August 1972), 174-184.

9. Corps of Engineers, *Post Flood Report,* Vol. II, 3.

10. Ibid.

11. Kneeland, *Playing Politics,* 15.

12. *e.g. Agnes Revisited.*

13. Notably: Frey, *The Flood of '72.*

14. Linda Stallone, *The Flood that Came to Grandma's House.* Dallas, Pennsylvania: Upshur Press,1991.

15. Nickerson, *It Sprinkled, it Rained, and it Poured*

16. Alan Rosenburg, excerpted from *Agnes,* Corps of Engineers, *Post Flood Report,* Vol II, frontispiece.

17. Community Collaborative Rain, Hail and Snow Network (CoCoRahs), accessed January 30, 2022, https://www.cocorahs.org/Content.aspx?page=aboutus.

18. "Military Patriarch Townend dies at 90," *Times-Leader,* Wilkes-Barre, Pennsylvania, Nov. 26, 2001, 3A.

BIBLIOGRAPHY

Accursia, Sister M. "Polish Miners in Luzerne County, Pennsylvania." *Polish American Studies*, Vol. III, No. 1-2. New Britain, Connecticut: Polish American Historical Association, 1946.

After Agnes: A Triumph Over Destruction. Wilkes-Barre, Pennsylvania: The Times-Leader, 1982.

Agnes ... And the Great Flood of '72. Philadelphia, Pennsylvania: The Sunday Bulletin, 1972.

Agnes Revisited. Wilkes-Barre, Pennsylvania: The Times-Leader, June 18, 1992.

Andersen, A.M., W.J. Davis, M.P. Lynch and J.R. Schubel. "Effects of Hurricane Agnes on the Environment and Organisms of Chesapeake Bay: Early Findings and Recommendations," *Special Reports in Applied Marine Science and Ocean Engineering*, 29. Williamsburg, Virginia: Virginia Institute of Marine Science, College of William and Mary, 1973. Accessed Jan 7, 2021. https://doi.org/10.21220/V5M44V.

Arnold, Joseph L. *The Evolution of the 1936 Flood Control Act*. Fort Belvoir, Virginia: U. S. Army Corps of Engineers, Office of History, 1988.

Atkins, Herbert. *The Wyoming Valley Floods of 1936*. Wilkes-Barre, Pennsylvania: The Collins Press, 1936.

Bailey, J.F., J.L. Patterson, and J.L.H. Paulhus. *Hurricane Agnes Rainfall and Floods June-July 1972*. Professional Paper 924, U.S. Department of the Interior, U.S. Geological Survey. Washington, DC: U.S. Government Printing Office, 1975.

Bechtel, Stefan. *Roar of the Heavens: Surviving Hurricane Camille*. New York: Citadel Press, 2006.

Bellini, Joe and Bill Kappel. "Flood Analysis for the World-Record-Setting July 1942 'Smethport' Storm," *Journal of Dam Safety* 16, No. 3 (2019).

Berry, James. *The Richmond Flood*. Lubbuck, Texas: C.F. Boone, 1972.

Blanchard, Wayne. "Deadliest Natural Disasters and Large Loss-of-Live Events." Accessed June 23, 2021. http://www.usdeadlyevents.com/1972-June-18-26-hurricanetropical-storm-agnes-flooding-east-coast-esp-pa-138-140/.

Bradley, Bob, director. *Agnes: A Flood of Memories*. 1992. Wilkes-Barre, Pennsylvania: Northeast Television Investors and WBRE-TV.

Bragg, Meridith, producer. *Stories from the Flood of '72*. 2012. Corning, New York: Corning Museum of Glass. Accessed June 23, 2021. https://www.cmog.org/video/stories-flood-72-john-fox.

Byrne, Thomas E. "The Impossible Flood," *Chemung County Historical Journal*, Elmira, New York, 1972.

Carter, Janet M., Joyce E. Williamson and Ralph W. Teller. *The 1972 Black Hills-Rapid City Flood Revisited*. U.S. Department of the Interior, U.S. Geological Survey. Fact Sheet FS-037-02, 2007.

Caviedes, Cesar N. "El Nino 1972: It's Climatic, Ecological, Human and Economic Implications," *Geographical Review* 65, No. 4 (October. 1975).

Clemens, Chris: "How the Mount Morris Dam Saves Rochester." *Exploring Upstate*, September 12, 2018. Accessed June 23, 2021. https://exploringupstate.com/mt-morris-dam/.

"Commerce Honors NOAA Employees," *NOAA Magazine*. Rockville, Maryland: U.S. Department of Commerce, January 1974, No. 1.

Cook, Tom and Tom Breslin. "The Flood of '72 in Letchworth Park", Exploring Letchworth Park History, Accessed April 23, 2020. http://www.letchworthparkhistory.com/flood.html.

Copp, Marvin W. "A Day to Remember, June 23, 1972", *Floods of the Chemung Watershed 1794-1972*. Elmira Heights, New York: Golos Publ. Co., 1975.

Cupper, Dan. "The Agony of Agnes." *Aspire Magazine*, June 1992.

Darmer, Kenneth I. and Lloyd A. Wagner. *Flood of June, 1972, at Elmira, New York* U.S. Department of the Interior, U.S. Geological Survey. Hydrologic Investigations Atlas HA-518, 1973.

DeAngelis, Richard M. "Hurricane Agnes June 14-23, 1972." U.S. Department of Commerce, National Oceanic and Atmospheric Administration, Environmental Data Service, *Climatological Data, National Summary* 23, No. 6 (1972).

DeAngelis, Richard M. and William T. Hodge. "Preliminary Climatic Data Report: Hurricane Agnes, June 14-23, 1972." NOAA Technical Memorandum EDS NCC-1, U.S. Department of Commerce, National Oceanic and Atmospheric Administration, Environmental Data Service, August, 1972.

del Marmol, Juio Antonio. *The Havana Conspiracies – Rites of Passage of a Master Spy*. Bloomington, Indiana: Trafford Publishing, 2016.

DiMego, Geoffrey J. and L.F. Bosart, "The Transformation of Tropical Storm Agnes into an Extratropical Cyclone," *Monthly Weather Review* 110 (1982): 385-411.

Edwards, Charleen. *Museum Under Water*. Corning, New York: Corning Museum of Glass, 1977.

Engman, E.T., L.H. Parmele and W.J. Gburek. "Hydrologic Impact of Tropical Storm Agnes." *Journal of Hydrology* 22, Issues 1-2 (June 1974): 179-193.

"Flashback to when a hurricane similar to Harvey hit DC," *WUSA-9 News*, Washington, DC, August 28, 2017. Accessed Oct 14, 2019. https://www.wusa9.com/article/weather/flashback-to-when-a-hurricane-similar-to-harvey-hit-dc/468706991.

The Flood of 1972. Ellicott City, Maryland: The Times Newspapers, 1972.

The Flood of '72 – Elmira Remembers. Accessed Dec. 31, 2020. https://www.elmira-ny.com/flood/index.shtml.

Friedlander, Blaine P. "Hurricane Floyd was strong, but was no Agnes." *Cornell Chronicle* (Sept. 23, 1999).

Forty Fort Cemetery Association. *In Memoriam: June 23, 1972*. Memorial booklet published by the Association, October 31, 1988.

Frey, Brian, writer and producer. 2012. *The Flood of '72*. Binghamton, New York: WSKG Public Telecommunications Council. Accessed Sept. 28, 2021. https://video.wskg.org/video/wskg-public-telecommunications-agnes-flood-72/.

Glahn, Bryan. *Hurricane Agnes in the Wyoming Valley*. Charleston, South Carolina: Arcadia Publishing, 2017.

"The Great Flood of 1972, Wilkes-Barre is Dealt a Powerful Blow." Accessed February 12, 2022. https://petejoem.wordpress.com/2014/02/09/the-great-flood-of-1972-wilkes-barre-pa-is-dealt-a-powerful-blow/comment-page-1/?unapproved=319&moderation-hash=1ca0f72dad85d74b95c930afdf820596#comment-319.

Grover, Nathan C. *The Floods of 1936*. U.S. Department of the Interior, U.S. Geological Survey. Washington, DC: U.S Government Printing Office, 1937.

Groves, Maj. Gen. Richard H. "The Agnes Disaster," *The Military Engineer* 65, No. 423, (January-February 1973): 15-21.

Hagemeyer, Bartlett C. and Scott M. Spratt. "Thirty Years After Hurricane Agnes - The Forgotten Florida Tornado Disaster," Presented at the American Meteorology Society's 25th Conference on Hurricanes and Tropical Meteorology, San Diego, California, 2002.

Healey, David. *Great Storms of the Chesapeake*. Charleston, South Carolina: The History Press, 2012.

Horton, Tom. "Retrospective – The Damage Caused by Hurricane Agnes," *Washingtonian Magazine*, June 19, 2012.

House, Kirk W. *The 1972 Flood in New York's Southern Tier*. Charleston, South Carolina: Arcadia Publishing, 2012.

Howard, Judith A. and Ernest Zebrowski. *Category 5: The Story of Camille, Lessons Unlearned from America's Most Violent Hurricane*. Ann Arbor: University of Michigan Press, 2007.

Hoye, R. Nicholas, Michael R. Orr and Joseph A. Anastasi. *The Flood and the Community*. Corning, New York: Corning Glass Works, 1976.

Hudgins, James E. "Tropical cyclones affecting North Carolina since 1586: An Historical Perspective," U.S. Department of Commerce, National Oceanic and Atmospheric Administration, National Weather Service. National Weather Service Technical Memo (April 2000).

"Hurricane Agnes: The Most Costly Storm," *Weatherwise* 25, No. 4 (August 1972): 174-184.

"The IBM 1130 Computing System," Accessed June 30. 2021. https://www.ibm.com/ibm/history/exhibits/1130/1130_intro.html.

"James River Flood, 1972," *Uvadisasters*. Accessed August 20, 2021. https://uvadisasters.fandom.com/wiki/James_River_Flood,_1972.

Kachic, Albert. *Agnes's Legacy*, presented at Alert-FLOWS East Coast Users Group, Philadelphia, Pennsylvania, 2002.

Kiefer, Robert G., director. *Storm – Hurricane Agnes/Wilkes-Barre*. U.S. Department of Agriculture, Motion Picture Division, a Public Information Film presented by the Defense Civil Preparedness Agency, 1972.

Kittlemann, Allan H. *The Ellicott City Flood Mitigation Plan*. Columbia, Maryland: Howard County Council, 2018.

Kneeland, Timothy W. *Playing Politics with Natural Disaster: Hurricane Agnes, the 1972 Elections, and the Origins of FEMA*. Ithaca and London: Cornell University Press, 2020.

Krantz, David L. *The Trouble with Agnes*. Wilkes-Barre, Pennsylvania: Fowler, Dick & Walker, 1973.

Lamont, Paul and Scott Saggett, writers, director and producer. *Lake of Betrayal*. Castle Films, 2017. Accessed September 30, 2021. https://www.lakeofbetrayal. com/.

Landsea, Christopher W. "Hurricane Harvey's Rainfall and Global Warming," National Oceanic and Atmospheric Administration. Accessed June 9, 2021, *https://www.aoml.noaa.gov/hrd/Landsea/harvey-global-warming.pdf*.

LaVere, Rich, writer/producer. *Facing the Wall*. Elmira, New York: Laveremedia, 2010.

Lazarick, Len. "Agnes' Historic Rainfall Taught Watershed a Concrete Lesson," *MarylandReporter.com*, June 27, 2012. Accessed July 30, 2021. https:// marylandreporter.com/2012/06/27/agnes-historic-rainfall-taught-watershed-a-concrete-lesson/.

Luman F. Robison, "Floods in New York, 1972, with Special Reference to Tropical Storm Agnes." United States Geological Survey, Water Resources Division. Albany, New York, 1976, 11.

Luzerne County Flood Protection Authority. Accessed Dec. 31, 2020. http://www. lcfpa.org/history.html.

Martin, John H. *The Corning Flood: Museum Under Water*. Corning, New York: Corning Museum of Glass, 1977.

Mason, Robert R. Jr. and Benjamin A. Weiger. *Stream Gaging and Flood Forecasting: A Partnership of the U.S. Geological Survey and the National Weather Service*, U.S. Department of Commerce, National Oceanic and Atmospheric Administration and U.S. Department of the Interior, U.S. Geological Survey, August, 1995. Accessed June 20, 2021. https://pubs.usgs.gov/fs/1995/0209/report.pdf.

Mathews, Lisa. "Tropical Storm Lee: 29T Gallons of Water Create Flash Floods." *International Business Times*, Sept. 9, 2011.

Mussari, Anthony J. *Appointment With Disaster: The Swelling of the Flood*. Wilkes-Barre, Pennsylvania: Northeast Publishers, 1974.

Namias, Jerome. "Birth of Hurricane Agnes – Triggered by the Transequatorial Movement of a Mesoscale System into a Favorable Large-Scale Environment," *Journal of the American Meteorological Society* 101, Issue 2 (February 1973).

National Advisory Committee on Oceans and Atmosphere. *The Agnes Floods: A Post-Audit of the Effectiveness of the Storm and Flood Warning System of the National Oceanic and Atmospheric Administration*. Report to the Administrator of NOAA, Nov. 22, 1972.

National Trust for Historic Preservation. *Shockoe Bottom*. Accessed Sept. 28, 2021. https://savingplaces.org/places/shockoe-bottom.

Nelander, John. "Anniversary: Hurricane Agnes Battered Keys, Slammed Panhandle 46 Years Ago," *Florida Weather Watch*, Palm Beach, Florida, 2018. Accessed June 24, 2021. https://floridaweatherwatchblog.wordpress. com/2018/06/19/anniversary-hurricane-agnes-battered-keys-slammed-panhandle-46-years-ago/.

New York Department of Environmental Conservation. *Corning and Painted Post*

Flood Damage Reduction Project. Accessed Jan 10, 2020. https://www.dec.
ny.gov/docs/water_pdf/fcpprjcornpp.pdf.

New York Department of Environmental Conservation. *Gang Mills Flood Damage
Reduction Project.* Accessed Jan. 10, 2020. https://www.dec.ny.gov/docs/
water_pdf/fcpprjgngmls.pdf.

Nickerson, John and Marina, writers and performers. *It Sprinkled, it Rained,
and it Poured.* Elmira, New York: WELM Radio, 1972. Video by Erin Doane,
Chemung County Historical Society, 2012. Accessed April 9, 2020. https://
www.youtube.com/watch?v=q7Q9xHjcv_g.

Paulson, John, writer and director, and Michael English, writer and executive
producer. *Conowingo Dam: Power on the Susquehanna.* Maryland Public
Television, 2016.

Penn Central Post. *Special Issue: The Flood.* Philadelphia, August-September 1972.

"Pinar del Rio and its Inland Geography", *Radio Rebelde.* Accessed Sept. 30,
2021. https://www.radiorebelde.cu/english/news/pinar-rio-and-its-inland-
geography-20130619/.

"Principales Eventos Pluviales Sobre Cuba en el Periodico 1963-2006," *CubAgua*,
Havana, Cuba: Instituto Nacional de Recurso Hidraulicos (July 23, 2011).

Romanelli, Carl J.. *The Wrath of Agnes.* Wilkes-Barre, Pennsylvania: Media
Affiliates, Inc., 1972.

Ryan, Diane. *The Great Flood of '72 Revisited.* Almond (New York) Historical
Society, 2007. Accessed February 15, 2020. http://www.usgennet.org/usa/
ny/town/almond/Archives%20List/Flood72/Flood72Revisited.htm.

Schroath, Judy A. "Dammed if you Do, Damned if you Don't," *Chesapeake Bay
Magazine*, April/May, 2011.

Scottsville Museum. *Capturing our Heritage: Scottsville Floods.* Accessed Sept. 28,
2021. https://scottsvillemuseum.com/floods/home.html.

Shafer, Mary A. *Devastation on the Delaware: Stories and Images of the Deadly Flood of
1955.* Riegelsville, Pennsylvania: Word Forge Books, 2015.

Shank, William H. *Great Floods of Pennsylvania.* York, Pennsylvania: American
Canal and Transportation Center, 1972.

Simpson, Robert H. *Hurricane Pioneer: Memoirs of Bob Simpson.* Boston: American
Meteorological Society, 2015.

Simpson, Robert H. Interview by Edward Zipster, American Meteorological
Society, 1989. Accessed Sept. 28, 2021. https://opensky.ucar.edu/islandora/
object/archives%3A7642.

Simpson, Robert H. and Paul J. Hebert. *Atlantic Hurricane Season of 1972.*
U.S. Department of Commerce, National Oceanic and Atmospheric
Administration, National Weather Service, National Hurricane Center, Miami,
Florida, 1973.

Simpson, Robert H., Arnold Sugg, and staff. *Atlantic Hurricane Season of
1969.* U.S. Department of Commerce, National Oceanic and Atmospheric
Administration, National Weather Service, April, 1970.

Stallone, Linda. *The Flood that Came to Grandma's House.* Dallas, Pennsylvania:
Upshur Press, 1991.

Stranahan, Susan. *Susquehanna: River of Dreams*. Baltimore: The John Hopkins University Press, 1993.

United States Army Corps of Engineers. *Tropical Storm Agnes!* 1973.

United States Army Corps of Engineers, Baltimore District. *Tropical Storm Agnes, June 1972: Post Flood Report* (two volumes), 1974.

United States Army Corps of Engineers, Buffalo District. *History of Mount Morris Dam*. Accessed April 23, 2020. https://www.lrb.usace.army.mil/Missions/Recreation/Mount-Morris-Dam/Project-History/.

United States Congress. House, Committee on Banking, Finance and Urban Affairs, Subcommittee on Policy Research and Insurance. *Aftermath of Hurricane Agnes*. Serial No. 101-147. U.S. Government Printing Office, June 22, 1990.

United States Department of Commerce. *Gold Medal Citation*, October 1973.

United States Department of Commerce, Environmental Science Services Administration. *The Virginia Floods, August 19-22, 1969: A Report to the Administrator*, September, 1969.

United States Department of Commerce, National Oceanic and Atmospheric Administration. *Black Hills Flood of June 9, 1972*, Natural Disaster Survey Report 72-1, Rockville, Maryland, August, 1972.

United States Department of Commerce, National Oceanic and Atmospheric Administration. *Final Report of the Disaster Survey Team on the Events of Agnes*. Natural Disaster Survey Report 73-1, Rockville, Maryland, 1973.

United States Department of Commerce, National Oceanic and Atmospheric Administration. *Incident News:Schuylkill River Spill*, accessed Sept. 30, 2021, https://incidentnews.noaa.gov/incident/6214#!513897.

United States Department of Commerce, National Oceanic and Atmospheric Administration. *Mariners Weather Log* 16, No. 6 (June 1972).

United States Department of Commerce, National Oceanic and Atmospheric Administration, National Weather Service. *May 27th, 2018 Flooding at Ellicott City & Catonsville, MD.* July 2, 2018, accessed June 20, 2021, *ttps://www.weather.gov/lwx/EllicottCityFlood2018*.

United States Department of Commerce, National Oceanic and Atmospheric Administration, National Weather Service, Middle Atlantic River Forecast Center, State College, Pennsylvania, 2017. "Hurricane Agnes - The 45th Anniversary," Accessed August 30, 2021. https://www.weather.gov/ctp/Agnes.

United States Department of Commerce, National Oceanic and Atmospheric Administration, National Weather Service, Northeast River Forecast Center. "Hurricane Agnes and the Genesee River Flooding," Accessed April 18, 2020. https://www.weather.gov/nerfc/hf_june_1972 .

United States Department of Commerce, National Oceanic and Atmospheric Administration, National Weather Service, Office of Meteorological Operations. *Preliminary Reports on Hurricanes and Tropical Storms: Hurricane Agnes June 14-23, 1972*. Silver Spring, Maryland, 1972.

United States Department of Commerce, National Oceanic and Atmospheric Administration, Office of Response and Restoration. *Largest Oil Spills Affecting U.S. Waters Since 1969*. Accessed Sept. 27, 2021. https://response.restoration.

noaa.gov/oil-and-chemical-spills/oil-spills/largest-oil-spills-affecting-us-waters-1969.html.

United States Department of the Interior, National Water Information System. "Peak Streamflow for the Genesee River at Wellsville," accessed April 17, 2020. *https://nwis.waterdata.usgs.gov/usa/nwis/peak/?site_no=04221000*.

United States Department of the Interior, Geological Survey. *Stream Gauging and Flood Forecasting: A Partnership of the U.S. Geological Survey and the National Weather Service.* Accessed June 24, 2021. https://pubs.usgs.gov/fs/1995/0209/report.pdf..

United States Environmental Protection Agency, Division of Oil and Hazardous Materials. *On-Scene Coordinator's Report: Schuylkill Oil Spill II, June-October 1972.* March, 1974.

United States Environmental Protection Agency, Office of Water Program Operations. *Environmental Effects of Schuylkill Oil Spill II, 1975.*

"The Violent, Deadly Swath of Agnes," *Time Magazine.* July 3, 1972, 9.

Walker, Paul K. *The Corps Responds: A History of the Susquehanna Engineer District and Tropical Storm Agnes.* United States Army Corps of Engineers, Baltimore District, 1976.

Wilson, James W. "Evaluation of Precipitation Measurements with the WSR-57 Weather Radar," *Journal of Applied Meteorology* 3, No. 2 (1964): 164-174.

Zebrowski, Earnest, and Judith A. Howard. *Category 5 – The Story of Hurricane Camille.* Ann Arbor, Michigan: Univ. of Michigan Press, 2005.

In addition, the following newspapers are cited herein:

The Aegis, Harford County, Maryland

The Sun, Baltimore, Maryland

Citizens Voice, Wilkes-Barre, Pennsylvania

Daily Item, Sunbury, Pennsylvania

Daily News, Lebanon, Pennsylvania

Daily Reporter, Wellsville, New York

Daily Review, Towanda, Pennsylvania

Democrat and Chronicle, Rochester, New York

Fort Lauderdale News, Fort Lauderdale, Florida

The Gazette, Montreal, Quebec

Intelligencer-Journal, Lancaster, Pennsylvania

Johnstown Chronicle, Johnstown, Pennsyvlania

The Leader, Corning, New York

Miami Herald, Miami, Florida

Morning Call, Allentown, Pennsylvania

Morning Herald, Hagerstown, Maryland

Naples Daily News, Naples, Florida

News-Herald, Panama City, Florida

The News, Frederick, Maryland

News-Item, Shamokin, Pennsylvania

News Tribune, Fort Pierce, Florida

New York Times, New York, New York

Olean Times-Herald, Olean, New York

Ottawa Citizen, Ottawa, Ontario

Ottawa Journal, Ottawa, Ontario

Patriot-News, Harrisburg, Pennsylvania

Philadelphia Bulletin, Philadelphia, Pennsylvania

Philadelphia Daily News, Philadelphia, Pennsylvania

Philadelphia Inquirer, Philadelphia, Pennsylvania

Pittsburgh Press, Pittsburgh, Pennsylvania

Pocono Record, Stroudsburg, Pennsylvania

Post-Gazette, Pittsburgh, Pennsylvania

Pottstown Mercury, Pottstown, Pennsylvania

Press and Sun-Bulletin, Binghamton, New York

Rapid City Journal, Rapid City, South Dakota

Robesonian, Lumberton, North Carolina

The Express, Lock Haven, Pennsylvania

The Sentinel, Carlisle, Pennsylvania

The Star, Washington, DC

Star Gazette. Elmira, New York

St. Petersburg Times, St. Petersburg, Florida

Sunday Telegram, Elmira, New York

Tampa Tribune, Tampa, Florida

Times-Dispatch, Richmond, Virginia

Times-Leader, Wilkes-Barre, Pennsylvania

Times-Mirror and Observer, Warren, Pennsylvania

Today – Florida's Space Age Newspaper, Brevard County, Florida

Virginia-Pilot, Norfolk, Virginia

Washington City Paper, Washington, DC

Washington Post, Washington, DC

The Weekender, Wilkes-Barre, Pennsylvania

Wyoming Valley Observer, Wyoming, Pennsylvania

York Daily Record, York, Pennsylvania

About the Author

Gary Letcher was studying geology at the University of Maryland at the time of the Agnes floods, and pitched in at cleanup around his grandparents' home at Nanticoke, Pennsylvania. Mr. Letcher practiced environmental law nationwide for many years, and now lives at Lewes, Delaware, where he teaches geology and genealogy at the Lifelong Learning Institute.

Mr. Letcher also penned *Waterfalls of the Mid-Atlantic States: 200 Falls in Maryland, New Jersey, and Pennsylvania* (Countryman Press/W.W. Norton, 2005), and *A Paddler's Guide to the Delaware River* (Rutgers University Press, 1985, 1999, 2012).

Inquiries, comments and orders may be directed to the author at g.letcher@PennDelPress.com

---

A Paddler's Guide to the Delaware River

"Indispensable, the definitive guidebook"
Upper Delaware Council

"Like having a personal guide on board"
Amazon.com

"A treasure for all seasons"
New York Times